AGROFORESTRY – INDIAN PERSPECTIVE

Edited by
PROF. L.K. JHA
Head, Deptt. of Forestry
School of Environmental Science
NEHU, Mizoram Campus, Aizawl-796007

and

PROF. P.K. SEN SARMA
Emeritus Scientist
Centre for Social Forestry,
NEHU, Shillong-793014

A P H PUBLISHING CORPORATION
4435-36/7, ANSARI ROAD, DARYA GANJ
NEW DELHI-110 002

Published by
S.B. Nangia
A.P.H. Publishing Corporation
4435–36/7, Ansari Road, Darya Ganj,
New Delhi-110002
Phone: 011–23274050
e-mail: aphbooks@gmail.com

2026

Typeset by
Ideal Publishing Solutions
C-90, J.D. Cambridge School,
West Vinod Nagar, Delhi-110092

Printed at
DIVINE DIGITAL PRINTERS
Ansari road Daryaganj Delhi-110002

Dedicated to

Late Dr. Gobind Jha

(1954-1992)

Whose untimely death has snatched away a potential agroforester. He was an active scientist of All India Co-ordinated project of Agroforestry (I.C.A.R.), Birsa Agril. University, Ranchi (Bihar).

Preface

The current concept of agroforestry implies integrating forestry with agriculture or agriculture with forestry. The practice existed in the tropical and subtropical countries since long. 'Taungya' system has been in existence in Myanmar (Burma) since ages. The practice was/is followed in Indian forestry since the advent of scientific forestry in India. Current emphasis on agroforestry emanates from the fact that the land is finite and the growth of population of human kind and cattle follows a geometrical progression. Therefore, the need is to evolve an integrated land management system in which goods and services can be obtained from the same unit of land for meeting the needs of food, fuel, fodder and fibre for human being and cattles. Agroforestry is thus truely the scientific basis of alternate agriculture, as tree husbandry is the foster mother of crop husbandry. There are several benefits such as conservation of soil and its moisture, increasing ambient humidity of an area, improving soil fertility, shading field crops against the malefic effects of scorching sun and dessicating effects of hot winds, etc. that can accrue when trees are incorporated in the crop farming system.

Numerous publications including proceeding of various national and international conferences have enriched our knowledge on this new applied science. However, many of the assumptions for agroforestry systems still remain empirical due to lack of research based data generation.

The current book has been conceived keeping in view absence of authentic data on agroforestry in the Indian context. Further some aspects have been rather neglected so far. The chapter authors have taken adequate care and pains on these aspects. It is hoped that the chapters though diverse in contents would generate sufficient interest among readers and would assist in furthering accumulation and enrichment of knowledge on the agroforestry.

L.K. Jha
P.K. Sen Sarma.

Contents

List of Contributors

1. A. N. Yellapa Reddy, Conservator of Forests, Research and Utilization circle, Bangalore.
2. B.A. Vadiraj, Scientist (Soil Science), Indian Cardmon Research Institute (Spice Board), Maladumpara, Iduki, Kerala.
3. C.M. Sharma, Head, Deptt. of Forestry, H.N.B. Garhwal University, Srinagar, Garhwal (U.P.).
4. Dinesh Kumar, Deptt. of Silviculture and Agroforestry, Dr. Y.S. Parmar University of Horticulture and Forestry, Nauni, Solan (H.P.).
5. J.C. Kaushik, Associate Professor, Deptt. of Agroforestry, Haryana Agril. University, Hisar, Haryana.
6. K.K. Mehta, Principal Soil Scientist, Central Soil Salinity Research Institute, Karnal (Haryana).
7. K.P. Nath, Head, Deptt. of Education, NEHU, Mizoram Campus, Aizawl.
8. L.K. Jha, Professor and Head, Deptt. of Forestry, NEHU, Mizoram Campus, Aizawl.
9. L.N. Singh, Head, Deptt. of Psychology, NEHU, Mizoram Campus, Aizawl.
10. M.A. Mohsin, Professor and Dean, Faculty of Agriculture, Birsa Agril. University, Ranchi (Bihar).
11. N.K. Verma, Head, Deptt. of Forest, Utilization, Faculty of Froestry, B.A.U., Ranchi (Bihar).
12. P.K. Sen Sarma, Emeritus Scientist, Centre for Social Forestry, NEHU, Shillong, Meghalaya.
13. Punam Roy, Vill+P.O. Kampur, Dt. Supaul (Bihar).
14. P.S. Pathak, Principal Scientist, Indian Grassland and Fodder Research Institute, Jhansi (U.P.).
15. Ramesh Jha, Scientist (Agroforestry), Rajendra Agril. Univ., Pusha, Dt. Samastipur, Bihar.
16. R.P. Tiwari, Head, Deptt. of Geology, Pachhunga Univ. College, NEHU, Mizoram, Ajzawl 796007.
17. Renu Ranjan, Professor, Deptt. of Sociology, Magadh Mahila

Womens College, Gandhi Maidan, Patna (Bihar).

18. R.S. Bhandari, Scientist, Division of Forest Protection (Entomology), F.R.I. Dehra Dun (U.P.).

19. S.D. Bharadwaj, Associate Professor, Deptt. of Silviculture and Agroforestry, Dr. Y.S. Parmar University of Horticulture and Forestry, Solan (H.P.).

20. Sushil Kumar, Scientist, Division of Forest Protection (Entomology), F.R.I. Dehra Dun (U.P.).

21. S.P. Arya, Associate Professor, Deptt. of Silviculture and Agroforestry, Dr. Y.S. Parmar University of Horticulture and Forestry, Solan (H.P.).

22. S. Parameswarappa, Principal Chief Conservator of Forest, Karnataka, Bangalore.

23. S.S. Bisla, Deptt. of Agroforestry, Haryana Agril. University, Hisar, Haryana.

Chapter 1

What Is Agroforestry ?

L.K. Jha*

Cutting of forest and growing of crop is a very old practice all over the world. Traditionally and historically forestry and agriculture are looked upon as competitive and not complementary activities. The ecological system is currently under great stress owing to growing pressure of human and livestock population on the one hand and acute shortage of food, fuel and fodder on the other. This has forced the people to change their attitude towards the exploitation of the ecological system. In the present time, integration of forestry with agriculture can only solve the problem. Under the protective umbrella of trees, agriculture can prosper and it is only with prosperous agriculture that forests can survive. The recent emphasis on agroforestry i.e. a coexistence of farm and forests all over the world is the outcome of the change of people's attitude out of necessity (Srivastra, 1985).

To make the subject more clear to the readers, it is essential to define the word agroforestry. We now proceed to extend the use of definition of forestry first and then social forestry/agroforestry.

As we know "Forestry is the theory and practice of all that constitutes the creation, conservation and scientific management of the forest and the utilisation of their resources".

The word Scientific management of the forest used in the above definition, in order to have continuous production of renewable goods and services. So we can now define forestry as "the theory and practice of all that constitutes the creation, conservation and continuous production of goods and services". Srivastra (1987) went one step ahead and he defined forestry as the "Scientific management of forests for continuous production of renewable, socially most acceptable goods and such services as are not

**Deptt. of Forestry, NEHU, Aizawl-796007*

subject to the principles of exclusion". The renewable socially most acceptable goods are fire wood, timber, fodder, bamboo, fencing material, resin, lac, honey, other non wood products, wild animals, birds, insects etc. and community service rendered by the forests include watershed protection, flood control, recreation, climatic regulation, waste treatment, CO_2 absorption, O_2 production, wildlife habitat maintenance and preservation of environmental quality. Till 1968 involvement of local people was limited and foresters worked in isolation. Westoby defined Social Forestry first time in the XI-Commonwealth Forestry Congress (1968) as "Social Forestry is a Forestry which aims at producing flow of protection and recreation benefits for the community". It was a revolutionary step as it aimed at the establishment of happy forest society relationship.

National Commission on Agriculture in its report (1976) stressed the socio-economic importance of social-forestry in the rural community as well as in the management of forest resource. The N.C.A. gave a broad base to the term social forestry and described as follows :-

(A) Farm Forestry (Agroforestry)

(a) Raising rows of trees on bunds or boundaries of the fields and individual trees in private agricultural lands.

(B) Extension Forestry

(a) Mixed forestry comprising creation of fuel, fodder, fruit and wood lots in the village common lands, Govt. wastelands and Panchayat lands.

(b) Raising of shelter belts.

(c) Planting of trees on road, canal and rail sides.

(d) Reafforestation of degraded forest lands situated in and around human habitats which lost their floristic and faunistic glory because of anthropogenic factors and biotic interferences.

(e) Recreation forestry.

(f) Urban forestry.

This clearly indicates that social forestry is a vast subject, we can define it differently, depending on the objective of the programmes.

Tiwari (1983) stated that social forestry which is an umbrella term with several components may, therefore, be defined in the Indian context as the science and art of growing trees and other vegetation on all land available for the purpose mainly outside traditional forest areas, with intimate involvement of the people and more or less integrated with other operations, resulting in balanced and complimentary land use with a view to providing a wide range of goods and services to the individual as well as the society.

On the basis of above discussion we can formulate general definition of social forestry as "Plantation outside the traditional forest area (marginal, submarginal private land and community land) and in degraded forest land (provided above categories of land are not available) by the involvement of individuals below the poverty line or socially and economically depressed people, with a view to meeting their requirements in respect of crops, legumes, tubers, fuel, fodder, timber, etc.

It is apparent that social forestry also includes agroforestry or we can say that agroforestry is an integrated part of social forestry and hence in above programme the production of food, crops, legumes etc., are feasible.

In fact "agroforestry" is a new name for an old practice which is now discussed under heading social forestry in Indian Forest Policies.

Growing crops and trees together is not a new invention or concept. This practice in different forms is age old in all the countries.

The system of raising trees along with agricultural crops has been in vague in Forest Department for over 100 years (Taungya cultivation). The usual practice has been to lease out the land to the cultivators to raise cereal crop after the annual coupe has been felled and exploited.

It would be worth mentioning here that in the year 1937 "The Hindu", a national Newspaper of India included the below mentioned features, in its 6th November issue.

A scheme known as agroforestry, which was a combination of agriculture and forestry, was adopted at the meeting of the Central Fodder and Grazing committee of the Imperical Council of Agriculture Research (Present: Indian Council of Agricultural Research) which met at New Delhi. The committee passed the following resolutions :

"The committee strongly approves the scheme put by the United Provinces (Present : Uttar Pradesh) Forest Department for a preliminary survey to examine the possibilities of creating fuel and fodder plantations in the plains districts (forestry-cum-agriculture known as Taungya in the United Provinces) has proved in forest areas most beneficial to the cultivator as a means of supplying poles, firewood and fodder, as also giving good land for cultivation and at the same time allowing regulated forest development. The possibility of applying this system to lands remote from forests and private lands is now to be explored".

Some foresters are of opinion that relegating a part of the farm to wood is agroforestry. Such ideas have led some people to define agroforestry as the technique of combining agriculture and silviculture in time or space. Such a definition is not only inadequate but is also misleading. Having agricultural crop at one time and raising forest at another, or growing agriculture crop in one area and raising forests in another cannot be correctly termed agroforestry. The technique of raising agricultural crops in combination with forest crops in the same unit of land and at the same time so as to maximise production from land is agroforestry. It is a system of interplanting following the principle of multiple use of land resource (Sinha, 1985).

Agroforestry has been defined by several people across the world and the definitions remained unsettled. The following definitions are worthy of mention.

Torres (1983) stated that agroforestry can be defined as the deliberate combination of trees with crop plantation or pastures, or both, in an effort to optimize the use of accessible resources to satisfy the objectives of the producer in a sustainable way. Lundgrean and Raintree (1983) narrated that agroforestry is a collective name for land use systems and technologies where woody perennials (tree, shrubs, palms, bamboos, etc.) are deliberately used on the same land management unit as agriculture crops and/or animals, either on the same form of spatial arrangement or on temporal sequence. In agroforestry system their are both ecological and economical interactions between the different components.

This definition outlines the broad boundaries of agroforestry and the typical characteristics of such systems :-

* Agroforestry normally involves two or more species of plants (or plants and animals) at least one of which is a woody perennial;
* An agroforestry system has two or more outputs;
* The cycle of an agroforestry system is always more than one year;

* Even the most simple agroforestry system is more complex ecologically, than a monocropping system.

According to Walt (1989) "agroforestry" has two dimensions. One is the way of thinking. It means looking for ways to maximize the production from land using a mixture of wood fibre, wildlife, recreation and farm crops, both plant and animal, to make the biggest buck possible consistent with keeping the land productive. The second dimension echoes the first. It is the actual management of land according to this ethic.

Rao (1989) mentioned that in the broadest sense, term "agroforestry" encompasses any and all techniques that attempt to establish or maintain both forest/tree and agricultural production on the same piece of land. According to the more strictly scientific definition of agroforestry, all forms of agroforestry are characterised by :

* The deliberate growing of woody perennials on the same unit of land as agricultural crops and/or animals, either in some form of spatial mixture of temporal sequence;

* There must be a significant interaction (positive and/or negative) between the woody and non woody components of the system, either ecological and/or economical.

Agroforestry also aims at systematically developing land use systems and practices where the positive interaction between trees and crops is maximized. This seeks to achieve a more productive, sustainable and diversified output from the land than is possible with conventional monocropping system.

Westley (1990) very correctly pointed out that researchers might also define agroforestry practices differently depending on the focus of their work. For instance, an agronomist interested in the effects of nitrogen fixing trees on the surrounding vegetation inevitably looks at an agroforestry practice differently from an economist interested in the potential cash income derived from selling of poles, once those trees are harvested. Finally, there's the question of scale: We can look at agroforestry from the perspective of individual trees and plants, of fields, of farms, of local geographic units such as watersheds or of broad ecological regions.

Readers can go through the paper written by Andriesse (1978), King (1978 and 1979), Huxley (1983), Foley and Barnard (1984), Bentley (1985), Rocheleau and Raintree (1986), Labelle (1987), Gordon and Bently (1990), Young (1990) for more informations.

REFERENCES

Andriesse, J.P. (1978). From shifting cultivation to agroforestry or permanent agriculture, Proc; 50th symposium, *Tropical Agril. Bull.*, 303, Department of agricultural research, Amsterdam, pp. 35-43.

Bentley, W.R. (1985). *Agroforestry—A strategy for research and action in India,* Discussion paper No. 17, The Fod Foundation, N.Delhi, p. 10.

Foley, G. and Barnard, G. (1984). Farm and community forestry, *Tech. Rep.*, 3. Earthscan, London.

Ghotz, H.L. ed. (1987). *Agroforestry : realities, possibilities and potentials,* Drdrecht (The Netherlands) : Martinus Nijhoff in cooperation with ICRAF, pp. 1-227.

Gordon, J.C. and Bentley, W.R. (1990). *A hand book on the management of agroforestry Research,* Winrock International, U.S.A. and South Asia Books, pp. 1-72.

Huxley, P.A. (1983). The role of trees in agroforestry. Some comments, In (Huxley, P.A. ed.), *Plant research in agroforestry*, Nairobi, pp. 257-70.

Huxley, P.A. (1983). Comments on agroforestry classification with special reference to plant aspect. In (Huxley, P.A. ed.), *Plant research in agroforestry*, Nairobi, ICRAF, pp. 161-171.

King, K.F.S. (1978). *Agroforestry*, paper presented in the 50th Agril. conf. Amsterdam.

King, K.F.S. (1979). Keynote address to the National seminar on Agroforestry. India, delivered on May 16-18, Shillong (Unpublished).

Labelle, R. (1987). Agroforestry : General concept, early work and current initiatives — a review of the literature, pp. 1-62 (Cross Ref.)

Lundgrean, B. and Raintree, J.B. (1983). Sustained agroforestry. Agril. research for development potential and challenges in Asia, The Hague, ISNAR, pp. 1-25 (Cross Ref.).

National Commission on Agriculture (1976). Govt. of India, Part IX (Forestry).

Rao, Y.S. (1989). Why, What, How and Where agroforestry in the Asia pacific region, *Forest News (Tiger paper)*, pp. 1-9, Thailand.

Rocheleau, D.E. and Raintree, J.B. (1986). Agroforestry and the future of food production in developing countries, *Impact of Science on society,* 142: 127-141.

Sinha, B.N. (1985). Role of agroforestry in soil and water conservation, In (Muhammad, S. ed.), Proceeding, *Social forestry workshop,* B.A.U., pp. 90-99.

Srivastra, J.L. (1987). *Defining social forestry*, Ensine Printers, Gumla, pp. 1-10.

Srivastra, V.C. (1985). Agroforestry concept, progress, In (Muhammad, S. ed.), *Social forestry workshop proceeding*, B.A.U., Ranchi, pp. 97-105.

The Hindu (1937). A National Newspaper of India, Dated 6th November (Cross Ref.).

Tiwari, K.M. (1983). Social Forestry as an aid to environmental conservation, *Van Vigyan*, Vol. 121, No. 1 & 2.

Tiwari, K.M. (1983). *Social Forestry in India*, Natraj Publishers, Dehra Dun, pp. 1-285.

Torres, F. (1983). Agroforestry : Concepts and practices. In (Hockstra, D.A. and Kuguru, F.M. ed.), *Agroforestry system for small scale farmers,* Proc. ICRAF/BAT workshop, Nairobi Sept. 1982, ICRAF, Nairobi, pp. 27-42.

Torres, F. (1985). *Net work of generation of agroforestry technology in Africa*, ICRAF, working paper 31, Nairobi, Kenya, pp. 28.

Walt, H.R. (1989). We need agroforestry in a big way, workshop in *agroforestry in California : planning for the twenty first century,* Deptt. of Forestry and Resource Management University of California, Berkeley, pp. 1-3.

Westoby, J.C. (1968). Changing objectives of forest management address to the *Ninth Commonwealth Forestry Conference*, New Delhi, 1968.

Westly, B. Sidney. (1990). Defining agroforestry technologies, A.F. databank, *Agroforestry today*, Vol. 2(1), p. 21.

Young, A. (1990). Agroforestry, environment and sustainability, *Outlook on Agriculture*, 19 : 155-160.

(This paper is part of Book written by Prof. L.K. Jha. He has permitted to reproduce this portion in this edited volume).

Chapter 2

Prospects and Problems of Agroforestry— An Alternative Agriculture

P.K. Sen Sarma* and L.K. Jha**

Conceptual framework

The concept of agroforestry eschews the artificial dichotomy of agriculture and forestry and is the step in right direction to evolve an integrated land use pattern. Agroforestry is, therefore, aptly defined as "a sustainable land management system which increases the yield of land, combines the production of crops (including tree crops) or animals simultaneously or sequentially, on the same unit of land and applies the management practices of local farmers (Lundgrean and Raintree, 1983). It is thus more or less a generic term to include agrisilviculture, silvipastoral, hortisilvipastoral, silviapiculture, silvilac culture etc., systems. This can be done in two broad manners, integrating crop cultivation with the tree growing and/or tree growing with crop production. However, Food and Agriculture Organisation of United Nation (FAO) had recognised the first three systems as the immediate requirements for the successful agroforestry programmes.

In the agrisilviculture system, agriculture crops including legumes are grown alongwith tree crops in the agricultural land. But growing forest trees alongwith agricultural crops in forests for a temporary period is an old system known in forestry terminology as "taungya" and this should not be confused with the present concept of agrisilviculture system.

Silvipastoral system envisages land management under which both trees, grass or seasonal fodder are grown simultaneously to promote animal husbandry. This system has in recent years included integrating

** Centre for Social Forestry, North Eastern Hill University, Shillong-793014.*
*** Deptt. of Forestry, NEHU, Mizoram Campus, Aizawl-796007.*

growing of leaf fodder tree species especially in areas where cattle are stall-fed with nutritious tree leaf fodder.

Horti-silvi-pastoral system includes growing seasonal horticulture crops (vegetable crops specially) alongwith forest tree crops and grasses/ legumes.

Demographic consideration and land use pattern

The importance of agroforestry development can be realised if demographic considerations are taken into account. By 2000 A.D., the human population of India will touch 1 billion mark with a projected food grain demand reaching a staggering 240 million tonnes. The current production level being around 175 million tonnes, the gap requires to be bridged up is about 65 million tonnes with hardly a decade left. Equal increase will be the cattle population for which grazing land has not only reduced in area but also been degraded to the extent that in summer months these areas are nothing but exercise grounds for them.

In India, more than 80 per cent of the potential arable land is already under cultivation. This indicates the limited possibility of increasing the acreage of the present arable land. However, culturable wastelands and marginal and sub-marginal lands constitute about 3,55,000 ha in Chotanagpur Plateau alone. These areas can be put into productive use through various agroforestry systems. In addition, vast areas in this plateau is monocropped on account of very low areas under irrigation. Thus, various agroforestry systems, if practised on a scientific line, not only increase the potentiality of putting this vast areas under productive use but also increase the fertility status through legumes and Nitrogen fixing tree species. Most tribal farmers have small land holding which negates inputs procurement to improve the productivity. They have also restricted resource base in terms of cash. Agroforestry systems have built-in autoregenerative factors to improve the land productivity.

Demand and supply of fuelwood

According to the Working Group on the Energy Policy appointed by the Planning Commission (1979), the consumption of fuelwood in India was 133.1 million tonnes against a recorded supply of 19 million tonnes from forests in 1975-76. Present projection of fuelwood need is about 220.0 million tonnes against a improved supply of 33 million tonnes from both traditional forests and social forestry plantations. Thus, the gap is progressively increasing. The worst sufferer for the energy crisis is the poorest section of the population including tribals. It is accepted that most of the fuelwood is extracted from forests as head-loads by the villagers residing adjacent to the forests and this has already dwindled forest cover. The gap in fuelwood availability can be bridged only through proper development and management agroforestry systems.

Shifting cultivation

Traditional *jhum* or shifting cultivation are practised by several tribes in India. In this practise, the areas in forests earmarked for *jhum* cultivation is clear-felled, timber and fuelwood taken out and sold and the area is put under fire so as to enhance the fertility of the soil. Crops are grown for 1-2 years using the most primitive methods of growing agricultural crops and area is left for regeneration of original vegetation. In earlier years, the fallow period between two *jhums* was long enough to permit natural regeneration of vegetation. But on account of population pressure, the fallow period is now too reduced to permit growth of natural vegetation with the result that the land is progressively becoming infertile. This is compounded by heavy soil erosion losses. Singh and Singh (1980) observed soil erosion at 147 tonnes/ha in the second year *jhum* and only 30 tonnes/ha in abandoned *jhum* in Meghalaya with 50-60 per cent slopes. This is much more than the acceptable limit. The erosion ratio is higher in the upper layer than in the lower layer of soil. Tang *et al.* (1979) compared the soil losses from forest, tea plantation and vegetable gardens and observed a ratio of 1:20:30 or equivalent of 24.5:488:732 $m^3/km^2/yr$. This clearly shows that soil losses is much higher under the shallow-rooted vegetable cultivation than in deep-rooted forest cover.

In addition, surface run-off in shifting cultivation areas in quite significant. Boarthakur *et al.* (1978) have observed that surface run-off under shifting cultivation is 111.4 mm. Partial terracing and complete terracing have reduced it to 81.4 mm and 32.8 mm respectively. A strip of trees planted along contours serves the same purposes as terraces. In most tribal area in India, soil losses and surface run-off are now much higher on account of loss of forest cover. Thus, a strip of trees along contours and agroforestry practices would reduce both surface run-off and soil erosion. Imposition of legislative measures to check this disastrous method would provoke resistance from tribal people who have been practising this system for centuries. Thus, there is a vast scope and potential to wean away the tribals from this autodestructive system through demonstration of appropriate agroforestry systems which will not be alien to their tradition cultural milieu.

Potential of agroforestry

The aim of agroforestry system is to optimise positive interactions between various biological components like tree/shrubs and crops/animals and between these components and the physical environment so as to obtain higher, more diversified and more sustainable production system from the available land resources than is possible with other form of land use under the prevailing ecological and socio-economic conditions (Lundgrean and Raintree 1983). Agroforestry approach to land development brings about the potential role of trees/shrubs to alleviate some of the major physical and economical constraints facing marginal and sub-marginal farmers in various tribal areas. On these lands, deliberate use of woody perennials, when property integrated in the land use system, restores/enhances both productivity and sustainability of land. This benefit is particularly essential to tribals for whom availability of capital and inputs is low and restricted. Agroforestry systems have built-in mechanism to enhance production of organic matter in order to maintain and enhance soil fertility, reduce soil erosion, maintain water balance, ultimately creating a favourable micro-climate. In the tribal areas, agroforestry systems are more relevant in view of lack of communications and markets inputs and cash. This will enable the tribal to produce most of the basic needs such as food, fodder, fuel and small timbers for shelter from a limited land area. Tribals are traditionally bound with forestry which they adore and worship. Therefore, this technology is not alien to their cultural milieu. Even the agronomist in several tropical parts are increasingly realising that only feasible long term approach for sustainable land use pattern is introduction of tree based land use system which the agroforestry aims at.

Benefits from agroforestry system

Agroforestry systems have both shortterm and long term benefits. Among the short term benefits, reduction of soil erosion, surface run-off and prevention of nutrient loss are more important. Soil erosion is less in land having deep rooted tree crop and more in shallow rooted agricultural crops. Reduction in surface run-off is achieved through tree crop planting. Use of multipurpose tree species in the cropping system is beneficial in improvement of soil nutrient status. Agroforestry systems also bring about less solar radiation, low soil temperature, improved soil organic matter, soil pH, lesser incidence of pests and diseases of plants etc. In forests, the canopy of tree species tends to reduce soil temperature and solar radiation reaching the forest floor. Agroforestry systems mimics these protective roles of forests. Favourable soil temperature and solar radiation improve the soil microbial activities, decomposition and nutrient cycling (Ewel *et al.* 1982). Legumes when included in the agroforestry system improves nitrogen availability of soil. If nitrogen fixing tree species (NFTs) are planted in the system, soil nitrogen increases considerably. Further, foliage of some NFTs contain considerable nitrogen which when mixed with soil improves the soil nitrogen status. Alley cropping with NFTs has proved this during experiments. It is known that soil organic matter is one of the factors affecting the soil structure which can account for the granulous nature of the soil. Further, shade of trees in the agroforestry systems improves soil micro-environment. This attracts beneficial organism such as earthworms whose activities would improve soil structure.

Pests and diseases are high in monoculture field and forest crops and much less in mixed crops. The ecological principles governing this is that the presence of other non-host crops check build-up of pest population and pathogen inocula.

By virtue of their protective and ameliorative roles, the immediate benefits from agroforestry systems as discussed above are followed by long term benefits such as increased crop productivity, sustainability of crop production through improved nutrient cycling, improved socio-economic conditions and health status of farmers. For a subsistence farmer, agroforestry is designed to increase and sustain crop production coupled with improved socio-economic conditions, nutrition and health status of farmer's family, multipurpose land use and improved environmental conservations.

As agroforestry practise include short term cash crops and long term tree crops, the systems are expected to ensure steady income for subsistence tribal farmers. Short term crops provide him his immediate need of food, fodder, fuel etc., while tree crops when matured would give him financial benefit in couple of years later. Sustained crop productivity is also ensured in the agroforestry practices. Pimental *et al.* (1983) have reported much higher yield of agricultural crop when grown along with tree crops. Agroforestry practices also improve their socio-economic status by way of his role in improving the fragile ecosystem of tribal areas and removing the stigma of the shifting cultivators. Availability of nutritious food (including fruits) would definitely improve health status of tribal farmer and his family.

Suitability characteristics of tree species for agroforestry

Tree species intended to be grown in conjunction with crops should *inter alia* have the following characteristics (King, 1978) :

(a) amenability to early escapement.

(b) Self-pruning property or capacity to withstand heavy artificial pruning.

(c) Low crown bole diameter ratio.

(d) Light branching.

(e) Tolerance to side-shading.

(f) Appropriate phyllotaxi for penetration of light to ground.

(g) Litter fall and litter decomposition rate having positive effect on soil.

(h) Absence of competition at root-zone level.

(i) Sufficient nutrient pumping.

(j) Amenability to silvicultural practices.

To summarise, crown architecture, morphology, phenology, root distribution, root spreading and activity of woody perennial must be taken into consideration while selecting tree species for incorporation in agricultural system.

Problems and constraints

The development of agroforestry systems and technologies, at several stages and levels, requires an integrated and multidisciplinary approach for which existing institutions and organisations are hardly equipped. Education at both technical and professional levels takes place along traditional disciplines like forestry, agriculture, animal husbandry etc. Research institutions, by and large, are compartmentalised in the similar manner. Even at government and administrative levels rigid compartmentalisation exists. In India, both at the Centre and State level, forestry and agriculture function under different ministries. It is also an irony that agroforestry development primarily rests with State Forest

Departments, efforts made by the Indian Council of Agricultural Research notwithstanding. Because of this dichotomy between agriculture and forestry, the village levels workers (VLWs) are normally trained in agricultural practices and are almost ignorant in forestry practices with the result that VLWs are technically ill-equipped to promote agroforestry adoption by assisting in transfer of technology in the agroforestry sector. This calls for delinking of social forestry wing of which agroforestry is an important component from the Forest Department in each State.

Development of agroforestry in tribal areas is also seriously handicapped on account of lack of institutional financing for agroforestry, though such facilities exist in case of agriculture, animal husbandry and fisheries. Even International aid agencies allocate financial aids for agroforestry through forest departments. Consequently, separate international research and development funds for agroforestry are rather scarce.

Another impediment in agroforestry development, both in tribal and nontribal areas, is the inherent constraints in the technology transfer at village level. Extension network is relatively well established in the agriculture and animal husbandry sectors but is non-existing in the forestry sector. Even extension education does not form an important curricular activity at the training centres of state and central government for forests guards, foresters, Forest Range Officers and higher executives. Lack of communication facilities like all weather road etc., in tribal areas even for those who are willing to transfer the technology to the underprivileged tribals. At the University level, the situation is not better either.

Since agroforestry is a comparatively new scientific discipline, the motivation of farmers to resort to modern agroforestry practices is another difficult task. A separate organisation is needed to convey to the farmers multifaceted benefits that agroforestry would bring through establishing a network of demonstration plots, adequate publicity literature etc.

Conclusion

From the discussion envisioned in earlier paragraphs, the multiple benefits of agroforestry systems are now well accepted. With the progressive increase in human and animal population and limited land area, there is hardly any alternative to agroforestry systems in order to meet the increasing requirement of food, fuel, fodder and animal products and timber for human kind and his herbivorous animals and for the environmental conservation and ecological security. But unless the constraints as elaborated above are removed, progressive acceptance of the systems will be restricted in nature. There is, therefore, an urgent need to introduce separate education and training for development, management and extension of the agroforestry systems at different levels. Indian Council of Agricultural Research is ideally suited to do this because of the existence of a network of Research Institutes and Agricultural Universities with the ICAR. As the systems are multidisciplinary, it is imperative to establish a separate wing for agroforestry development in the Directorate of Agriculture in each State. The wing should comprise professionals in forestry, agronomy, horticulture, animal husbandry etc., both at the officer and subordinate levels. The systems, by its very diverse nature, cannot be handled by any of the traditional disciplines and departments as exist today.

REFERENCES

Borthakur, P. et al. (1978). Ecological consequences of traditional land uses in loss of productive soil. *In South east Asia Regional Symp. on problems of Erosion and Sedimentation,* Bangkok (Eds. T. Tingsanthal and H. Eggers) : 27-28.

Ewel, J.S., Gleissman, M. Amador, Benedict, C., Berish, R., Bermudez, R., Broun, R., Martinez, A Miranda, R. and Price, N. (1982). Leaf area, light transmission, roots and leaf damage in nine tropical plant communities. *Agroecosystems 7;* 305-326.

King, K.F.S. (1978). 50th *Agric. Conf., Paper,* Amsterdam.

Lundgrean, B. and Raintree, J.B. (1983) : In *Agricultural Research for Development Potentials and Challenges in Asia,* ISNAR, The Hague 1-25.

Pimental, D.P., G. Benardi and S. Fast (1983). Energy efficiency of farming systems : Organic and conventional agriculture *Agric. Ecosystem and Environ., 9:* 359-372.

Singh, A and Singh, M. (1980). Effect of various stages of shifting cultivation on soil erosion from steep hill slopes. *Indian Forester, 106:* 116-121.

Tang, H., Monararam, N. and Blacke, G. (1979). The status of hydrological studies at the Forest Research Institute, Kepong. *Malaysian Forester, 42:* 108-114.

Chapter 3

Population, Resource Availability, Environment and Agroforestry Intervention

L.K. Jha and K.P. Nath***

A study of the present demographic situation reveals that there has been an increasing trend in the poverty of our country. The number of poor living below the poverty line has increased from 176 million in 1961 (40% of the then population of 439 million) to 292 million in 1988 (37% of the then population of 790 million). It is apparent that after this decade our population will go up by above 170 million. By the time this happens, it is likely that the number of people below poverty line will exceed 500 million, constituting thereby at least 50 per cent of the population (Vohra, 1991).

Per capita cultivable land has declined from 0.48 hectare in 1951 to 0.26 hectare in 1981. It is apprehended that by the year 2000 A.D., it will go further down to around 0.11 hectare per capita (Choudhary, 1989).

The hopeful side of the picture is that the production level of food grains has gone up from 50 million tonnes at the time of independence to 180 million tonnes at the present moment. This has been possible because of the use of high yielding variety of seeds, chemical fertilizers and insecticides. The production of rice has increased by 5.34 million tonnes and yield per hectare by 200 kg (Siddiq, 1988). In 1988-89 maize growing area went up to 5.95 million hectares and the yield was 8.33 million tonnes (Govila, 1990). The area under millet, called the poor man's ideal food, rose by 91.8 per cent but the production dropped by 82.6 per cent during 1985-1987. This was mainly due to drought (Harinarayana, 1988). Wheat is grown in 23.07 million hectare and production went up to 12.26 million tonnes. The yield per hectare increased from 913 kg to 2,032 kg (Tandon, 1988).

**Department of Forestry, NEHU, Mizoram Campus, Aizawl.*
***Deptt. of Education, NEHU, Mizoram Campus, Aizawl.*

The majority of the people in India gets protein from pulses. Farmers have given up growing pulses in irrigated farms to get maximum production from other cereals (Kanwar, 1988). India spent about Rs. 7,000 crores between 1976-77 and 1986-87 on edible oil import. To reduce the drain of foreign exchange and dependency on other countries, it was decided to grow 26 million tonnes of oil seeds to yield 80 lakh tonnes of oil (Rao, 1988). Presently we are cultivating sugarcane in the largest area of world. In 1986-87, we have produced 85.10 lakh tonnes of sugar. Rao (1988) pointed out that the need of the hour is the stabilization of sugarcane and sugar production to meet the challenge of the 21st century and compete with the international market. For an estimated one billion population by 2000 A.D. with a per capita consumption of 15 kg, 300 million tonnes of cane and 150 lakh tonnes of sugar have to be produced with a sizeable quantity to spare for export. Fruits and vegetables combinedly cover about 5.2 million hectares out of which perennial horticulture crops account for 4.4 million hectares with an annual production of 54 million tonnes (Singh, 1988).

Livestock is an important component of our rural economy. It provides additional income to the small and marginal farmers. India is blessed with a huge livestock population of 19.2 crores of cattle, 7.0 crores of buffaloes, 9.5 crores of goats, 4.9 crores of sheep and 21 crores of poultry birds as per the 1982 livestock census. This number must have increased significantly by now. It is expected that about 125 million bullock will be required to provide 60,000 MW of animal power by 2000 A.D. and milk requirement will be around 76 million tonnes (Annon, 1991). Alongside cattle, the farming of pig, goat and poultry is essential to meet the requirement of the nation and improve the economic condition of farmers.

The Committee on fodder and grass appointed by the Government of India in 1985 estimated the availability of only 229 million tonnes of the same as against the requirement of 672 million tonnes. It has been estimated that by the end of 2000 A.D., approximately 949 million tonnes of dry fodder and 1200 tonnes of green fodder will be required to rear our cattle population.

Our third basic requirement is energy. The Advisory Board of Energy (1985) reported that the present requirement of energy from non-conventional sources like fuelwood, cowdung cake and agricultural waste is around 229 million tonnes. By 2000 A.D. it will be around 300 million tonnes (Table 3.1 & 3.2). Martin (1990) stated that the shortage of fossil fuels have constrained economic development and prevented urban households and industries from switching away from wood and other biomass fuels. Worst still, deforestation and desertification are threatening traditional energy supplies thus starving the rural sector of even biomass fuels at the very time when more efficient sources of energy are needed for rural development. Because of the scarcity of energy inputs, productivity and income generation are poor and the rural people have no option but to meet their subsistence needs through non-commercial energy source, which has resulted in large-scale destruction of environment (Chopra, 1988).

Table 3.1 :Sectoral breakup of energy requirements by 2000 A.D.

Item	*Household*	*Industry*	*Transport*	*Agriculture*
Commercial				
Electricity				
(Billion KW)	81.40	249.00	8.20	40.5
Coal (M.Tonne)	14.00	261.00	7.81	-
Oil (M.Tonne)	17.70	12.00	29.90	8.1
Non-Commercial				
Fuelwood (M.Tonne)	191.60	-	-	-
Dung Cake (M.Tonne)	105.00			
Veg. Waste (M.Tonne)	59.00			

Source : Report of the Advisory Board on Energy, 1985.

Table 3.2 :Present and projected non-commercial Energy Consumption in rural India

Item	*Present*	*Projection for 2000 A.D. (MTCR Units)*
Fuelwood	95	182
Dungcake	96	42
Agricultural waste	38	56
Total	229	280

Source : Report on Advisory Board on Energy, 1985.

It is apparent that extensive deforestation, indiscriminate use of land, burning of fuelwood, application of insecticides and fertilizer, etc., have created global atmospheric change. Pearman (1990) has stated that the planet earth has a mean surface temperature of 15°C, about 30°C above the temperature that would exist if there were no infra-red absorbing gas and green house gases in the atmosphere. These gases include Water Vapour, Carbondioxide (CO_2), Methane (CH_4) and Nitrous Oxide (N_2O). The global activity of man are changing the concentrations of these naturally existing gases and adding additional green house gases, which are entirely man-made. Increased evaporation and increased level of atmospheric water vapour are likely to increased cloud cover. Clouds are known to have a cooling effect on the atmosphere due to their reflecting away the incoming sunlight and thus counteract to a certain extent the effect of green house warming. At this stage it is not clear which of these two processes would dominate (Ramanathan, *et al*, 1989). It is argued that if the current trends of the emission of non-carbondioxide trace gases continue, nearly 50 per cent of the projected warming of the planet would arise from the non-carbondioxide greenhouse gases. It is considered essential to reduce the emission of non-carbondioxide gases by 50 per cent and carbondioxide by 66 per cent to ensure that the earth's warming rate is maintained at 01.°C per decade (Ramasami *et al*, 1989).

Great climatic change and erosion of fertile top soil of the earth due to large-scale deforestation will affect the production of foodgrains. Unless precautionary measures are taken right from now food-scarcity will cause concern as we will enter the 21st century.

The following are considered the areas of relevance to tackle the problem:

A check in the growth of human and livestock population

To maintain ecological balance, human as well as livestock population should be checked so that desired produce in the form of food, fodder and fuel can be obtained from natural resources on a sustained basis. Both human and livestock population must not increase beyond the carrying capacity. As for the livestock it should be borne in mind that a fewer healthy and productive breed is always better than many sickly and unproductive ones.

Evolving sound land use policies

Food security for all people can be ensured by proper land use planning and policies. Agroclimatic regional planning (ACRP) was initiated by the planning commission in 1988. Fifteen agroclimatic zones were formed on the basis of soil and water resources, distribution of rainfall and temperature. A zonal planning team (ZPT) was set up for each zone under the chairmanship of the Vice-Chancellor of an agricultural university in the region and involving experts, senior officials of the states concerned, farmers, representatives of banks and some voluntary groups as members. The ZPT's have submitted reports and made recommendations on the cropping pattern, land and water use management, dairying, fisheries, forestry and other agricultural allied enterprises at the sub-regional level.

The analysis of the potential and constraints taken up by the ZPT's indicates that major focus on the following types of schemes will have a definite impact on the output and productivity of various crops in different regions of the country.

(a) A target of 40 million hectares for integrated development programmes.

(b) A target of over 5 million hectares for restoration for tank irrigation.

(c) Preparation of operational water delivery system for 50 per cent of the area under canal irrigation and of rehabilitation of a quarter of the area under the canal command.

(d) Operational plan for dairying and fisheries.

(e) Operational plan for high-value-low-volume crops such as horticulture, spices and plantation.

The ACRP strategies have been reflected in the plan documents of the various states and this integration is significant in making the ACRP a timely, relevant and practical exercise. The strategies are based on the need for maximum efficiency in resource use in the long term setting of sustainable agriculture (Singh, 1990). There is need for successful implementation of the above programmes. Sericulture, lac-culture and apiculture need also be included in the operational plan.

Development of appropriate cropping system

Development of appropriate cropping system like agroforestry technology to different climatic zone will help in reducing the harmful global atmospheric changes. Tree planting has recently been suggested as a method of countering the green house effect. However, the sheer magnitude of the task at hand needs to be kept in mind. Ten billion of trees if planted today will occupy 10 million hectares of land and will take up enough carbondioxide from the atmosphere to reduce the world's annual carbondioxide emission from burning fossil fuels by about 1 per cent (Marland, 1988).

Thus, in India planting of trees in farmland under the agroforestry system will help in achieving the target of tree plantation and will help in countering the effect of atmospheric carbondioxide and protect land from degradation and maintain ecological balance.

The key features of the agroforestry system are:

1. Fruitful utilisation of uncultivable wasteland, farmland, use of crop rotation etc., and planting of NFT's to reduce the need for purchasing fertilizer.
2. Production of resources to meet local demands as well as national needs by adopting silvipastoral, agrisilvipastoral and multicropping systems.
3. Provision of animal feeds on a sustained basis.
4. Suppressing of weed growth.
5. Recycling of available resources.
6. Economic rehabilitation of poor farmers by creating perpetual income opportunities through quick and sustained returns from agricultural crops, fruits and vegetables.

It is essential to mention that due to rise in atmospheric temperature, the yield from present day variety of crops may be reduced. We should, therefore, continue our research on identifying temperature tolerant cereals, pulses, oil seeds, etc., and on developing high yielding hybrids right from now.

Nutrient management

Nutrient management under some important cropping system was studied under various ICAR projects. Our needs can be met by adopting agroforestry system such as the alley cropping and the like. This system can ensure supply of food, fodder, fruit, milk and meat on a sustained basis in the 21st century.

To increase productivity judicious use of green leaf manure, fertiliser, micronutrients and their availability at village level is essential. We have to train farmers about the intensity, time of harvesting of green foliage and methods of use, so that can improve productivity with minimum use of fertilizers.

Hybrid varieties

It is advisable to use hybrid seeds to increase the yield of crops substantially. Under agroforestry system research should be conducted to study the potential of genetically superior high yielding varieties in agroclimatic zones. Extension workers should encourage farmers to use only available hybrid varieties. There is need for supplying of seeds to the farmers in the village by the Government.

Irrigation

In order to get desired produce today as well as in the 21st century an appropriate system of water management is essential. It is necessary to study the quantity of water required to produce desired yield in different agroclimatic zone. We can tackle water problem by adopting below mentioned irrigation strategies:

* Controlling of irrigation so as to stop over-irrigation by educating farmers about its evil effects.
* Conjuctive use of surface as well as ground water.
* Use of sprinkler system and drip irrigation in area having scarce ground water resource.
* Creation of chief water source like dug well along with recharging tank.
* On-farm rain management in dry farming technology.

Data on rainfall probabilities

This will help in calculating the length of growing season and variability. Farmers can accordingly adopt susceptible cropping patterns. A concerted and well-coordinated effort between agriculturists and meteorologists is needed to harness the sophisticated tools and data now available and the data which can be generated to benefit the individual farmers and to stabilise and boost the country's improving crop and food production (Vironani, 1988).

On-farm trial

In many instances, the results of the experimental station remain unused or newly evolved production technology is not accepted by farmers. Sometimes technology is transferred in the farmer's field without considering the biophysical and psycho-socio-economic problems of the farmers. Farmers never accept imposed technology. Under the above circumstances, the impact remains for a short period. Farmers may accept any new innovations, provided the extension workers or the scientists works along with the farmers to examine the requirements of the latter and modify the tested technology to suit their special needs. Once the scientists or the extension workers start working in collaboration with the farmers, the latest technology will be accepted by the farmer as a continuing feature on a long term basis. It will be a wise step to conduct experiment on-farm in collaboration with farmers.

Effective Extensive Wing

There is need for an active and effective extension Department in each agro-climatic zone. By creating awareness and developing of confidence the latest technology may create desired impact in the farmers. In all extension programmes emphasis should be given on active participation of farmers in implementation as well as evaluation. There must be adequate budgetary allocation to support extension work at the village level to create sustained interest among farmers towards adopting any new technology in farming. There can be also suitable training programme to train farmers as well as functionaries at different levels.

Farm price policy

In order to inculcate interest among farmers to increase productivity, the Government should change present farm price policy. There is need for an efficient price support mechanism to create incentive among farmers to grow more food crops. The price policy influences the farmers to adopt improved technology in farming in order to increase productivity. Side by side there is need for creating efficient marketing system. Facility should be created so that farmers can sell their produce in bulk at their door step to the Government in competitive price.

Socio-economic transformation of village

Most of the ongoing anti-poverty programmes have benefited those having land. The majority of the people lying below the poverty line are agricultural labourers and marginal farmers. They are not effectively integrated into the ongoing programmes. Still they are facing problems of poverty and presently more than 50 per cent households in rural areas are still below the poverty line.

It is an accepted fact that rural transformation is possible only when employment is generated in the villages, when farmers will be provided with packages of technology, proper guidance and short term credits. Any rural transformation programme should aim at total socio-economic transformation of the villagers. The success of the programme depends upon the collaboration of the Government, Agricultural University, Voluntary Agencies and full participation of the target group.

Steps involved in rural transformation on a model basis (Jha, 1991)

* Identification of village under different socio-geographical region.
* Identification of beneficiaries (target group).
* Field exercise by the extension worker in agro-ecosystem analysis.

This will be based upon the visits in the village, formal interviews with the farmers and collection of information related to land type, soil, crops and trees grown, livestock, lacculture and sericulture practised and other problems of the target group. The extension worker will prepare maps and transects of land type and problem should be listed below the baseline of the transect. Questions should be asked in the field. Participants will explain in details about the various related matters.

* Identification of biophysical causes. System diagrams must be drawn after knowing socio-economic and biophysical causes.

The findings of the above study will provide empirical foundations to formulate programmes that will minimise poverty in the village. In the present situation we should always try to introduce cropping system which will provide products of agriculture (cereals, pulses, vegetables and oil), trees (fuel, fodder and fruits) and animal husbandry (meat, milk and egg).

Appropriate technology for agroforestry and pasture management and allied practices to suit the prevailing agro-climatic condition have been developed by ICAR and state agricultural universities. This system will fulfil the requirements of the rural poor and will improve their socio-economic condition. The most interesting point in favour of this system is that it improves fertility of land and gradually converts marginal land into viable economic unit.

A few suggestions

* The Government should allot land permanently to the landless participants so that they can be benefited along with participants having the basic asset of land.
* Village level cooperative society as per model provided in the flow chart (Fig. 3.1) should be constructed to harness maximum benefits.

The cooperative society will endeavour to create awareness among the farmers, motivate them to adopt latest technology and will coordinate technical experts and the extension workers for providing farmers the technical know-how to get produce from their farms on a sustained basis. Farmers will also get loan from the society for input and for purchasing cattle, pig, goat and poultry. This will assist them in raising their income and consequently their socio-economic condition.

Thus in the rural transformation programme farmers will get incentives through the cooperative society. The farmers can utilise the land and resources available to them by introducing agroforestry system. Under this system they should divide their land available for agriculture into smaller plots of suitable sizes and grow high density plantation in one plot for fuel wood and timber and fruit trees in another with spacing as suggested by the extension workers. In the intervening plots agricultural crops and vegetables can be grown. Hedge rows can be raised bordering each plot to protect the land from soil erosion. Farmers can put one plot for multicropping purpose. Farmers can also grow perennial flowers in

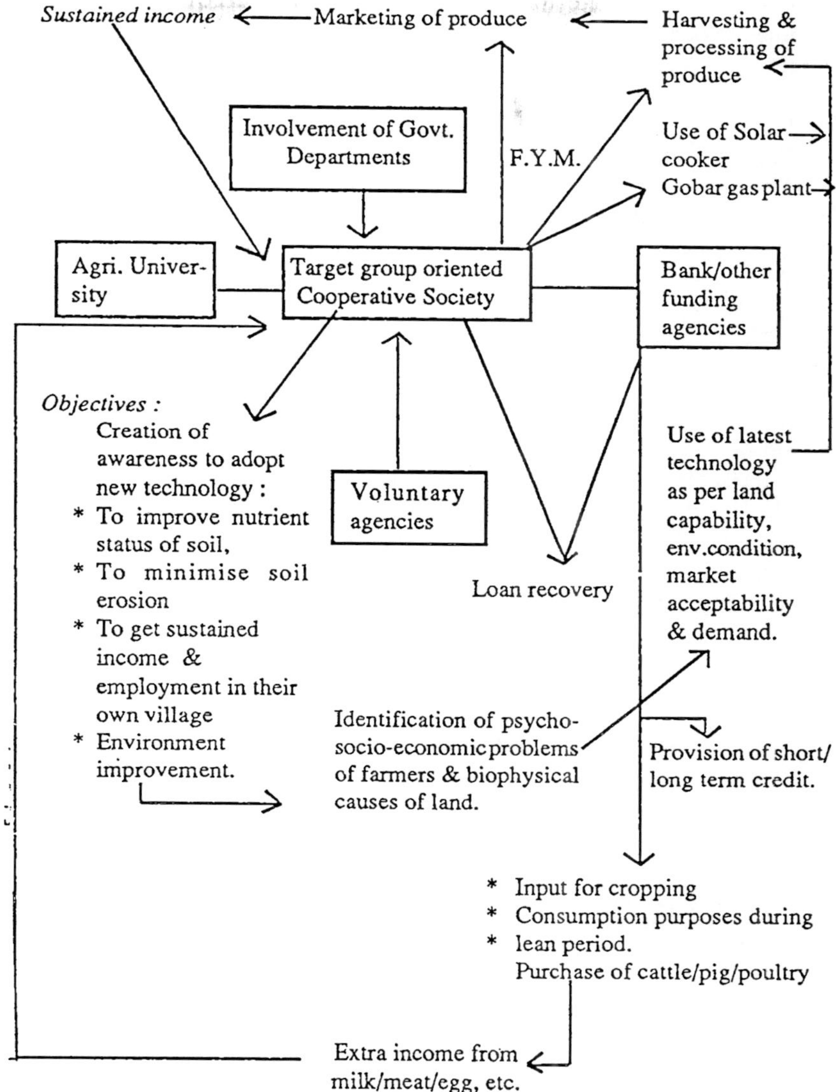

Fig. 3.1: Cooperative Society Flow Chart

one small area of his field and can practise apiculture. Likewise mulberry trees can also be grown for silk-worm rearing, which will supplement the farmers' income. Green foliage of the hedge rows can be utilised as green manure or as fodder. With the financial assistance from the Government the farmers can also develop side by side cattle, pig, goat or poultry farming. These will help augment their income. In the initial stage the farmers will manage expenses from the loan provided by the Government. The loan will be recovered in instalment through the cooperative society. The success of this programme depends upon the coordination between the agricultural, forestry, animal husbandry and other allied departments. The Cooperative Department can play the role or coordinator.

Seventy-seven per cent of India's population live in villages. Transformation of India cannot, therefore, be possible without transforming the rural farmers who constitute the bulk of India's population. It is on their prosperity that depends ultimately the prosperity of the nation. Our farmers must be made conscious of the role they are going to play in the coming years for the upliftment of the nation by producing more food and at the same time endeavouring to restore ecological balance in nature by planting more trees. They have to shoulder this dual responsibility. Our farmers will have to rise to the occasion and make India emerge as a peaceful, progressive, prosperous and powerful nation as she steps into the 21st century.

REFERENCES

Anon (1991). *India 1991.*, A Reference Annual, Publication Division, Ministry of Information and Broadcasting, Govt. of India, pp. 1-29.

Chopra, S.K. (1988). Rural Energy, Integrated Approach, *The Hindu, Survey of Indian Agriculture,* pp. 21-28.

Chowdhry, K. (1989). Poverty and Land Use Policies, *Forest News,* F.A.O. Regional Office from Asia and the Pacific as part of Tiger paper, pp. 5-10.

Govila, O.P. (1990). New Direction in Millets, *The Hindu, Survey of Indian Agriculture,* pp. 51-53.

Harinarayana, G. (1988). Millets: Growing role in Food Economy, *The Hindu, Survey of Indian Agriculture,* pp. 38-42.

Jha, L.K. (1991). Rehabilitation of Jhum land through Social Forestry, S.F.S.O., Shillong, pp. 1-10.

Marland, G. (1988). The Prospect of Solving the Carbondioxide problem through Global Reforestation, Washington, D.C., United States Department of Energy, TRO 39 (DOE/NBD-0082), pp. 1-66, cited (Noller, B.N. & Chadha, M.S., Ed. 1990) *Chemistry and Environment, Proceeding of Regional Symposium*, Brishane.

Martin, T.P. (1987), Foreword: Linking Energy with Survival, A Guide to Energy, Environment and Rural Womens Work, International Labour Office, Geneva, pp. 1-32.

Mishra, H.R. (1989). Rural Transformation in Tribal Bihar, (KVK Souvenir) Ranchi, pp. 28-32.

Pearman, G.I. (1990). The Green House Effect: Global and Australian Perspective, In (Noller, B.N. & Chadha M.S., Ed.) *Chemistry and Environment,* Commonwealth Science Council, pp. 37-39.

Ramanathan, V., Cess, R.D., Harrison E.F., Minnis P., Barkstrom, B.R., Ahmad, E. and Harlman, D. (1989). Cloud Radiative Forcing and Climatic Results from the Earth Radiation, Budget Experiment, *Science,* Vol-243, pp. 57-63, (Cited from Noller B.N. & Chadha M.S., Ed.) *Chemistry and Environment, Proceedings of Regional Symposium* (1990).

Ramasami, T., Chadha, M.S., and Thyagarajan, G. (1989). Chemistry and Environment, Commonwealth Science Council *Tec. Bull series* No. 261, pp. 1.48.

Rao, B.V. (1988). Poultry, A Neglected Sector, *The Hindu, Survey of Indian Agriculture,* pp. 193-194.

Siddiq, O.P. (1990). New Direction in Millet, *The Hindu, Survey of Indian Agriculture,* pp. 51-53.

Singh, H. (1988). Fruits: Vast Untrapped Potential, *The Hindu, Survey of Indian Agriculture,* pp. 121-124.

Singh, H. (1990). Agricultural Development: Towards a Strong Resource Base, *The Hindu, Survey of Indian Agriculture,* pp. 29-33.

Tandon. J.P. (1988). Wheat Target for 2000 A.D.: Overcoming Serious Constraints, *The Hindu, Survey of Indian Agriculture,* pp. 29-33.

Virmani, S.M. (1988). Rain Forecast as a Vital Tool, *The Hindu, Survey of Indian Agriculture,* pp. 235-336.

Vohra, B.B. (1991). Tackle Poverty on a War Footing, The Times of India, November 7, 6 pp.

Chapter 4

Multipurpose Trees and Shrubs for Agroforestry Systems

*P.S. Pathak**

Introduction

Destruction of tropical forests has been most rapid in the past fifty years due to heavy harvesting, systematic clearing and usurping forest lands to agriculture, hydroelectric projects, shifting cultivation etc. to meet the growing needs for food, industrial raw material and other developing projects. Rate of deforestation is more serious in the developing countries. About 11 million hectares forests are removed each year for various purposes and less than 10 per cent are returned to forest vegetation on a global basis (MacDicken and Sastry 1988).

Use of wood for domestic needs has been associated with the evolution of human civilization through the forests. In tropics and subtropics has climax forest vegetation. During development of settled agriculture some very useful woody species were needed for day to day use. Such species were, therefore, cultured out of necessity since they catered to the multipurpose needs of the rural habitat. With the fast decline in the area under forests and also its productivity, pressure has increased on these woody perennials stripping off the permanent woody cover leading to the non sustainable agroecosystems. On degraded soils and fragile ecosystems it has become all the more difficult to establish these species with the existing techniques.

In the past few decades potential of tree legumes has been well recognized for improving agricultural and silvicultural production especially in the third world countries (Felker and Bandursky 1979). Many species are now widely cultivated outside their country of origin

**Indian Grassland and Fodder Research Institute, Jhansi-284003*

and have found highest adoption in various farming systems (notable examples are *Leucaena leucocephala and Gliricidia sepium*).

Multipurpose trees and shrubs had been the basic theme of discussion in many National and International seminars during the past decade and various issues related to their use and introduction have been discussed.

What are MPTs ?

The multipurpose trees and shrubs could be defined as trees grown deliberately or kept and managed for preferably more than one intended use, usually economically motivated major products and or services in any multipurpose land use system, especially agroforestry system (Van Carlowitz and Burley 1984). Van Carlowitz (1984) further defined it as a tree that clearly constitutes an essential component of an agroforestry system or other multipurpose land use systems. Regardless of the number of its potential or actual uses, a multipurpose tree must have the capacity to contribute, in its function(s), in the system a substantial or a recognizable benefit to the sustainability of yields, to increase of the output and/or the reduction of inputs, and to improve the ecological stability of this system. Only a tree that is kept and maintained or introduced into an agroforestry system especially for one or more of these purposes qualifies as a multipurpose tree.

Thus, multipurpose trees and shrubs should provide one or more significant products or services to the people in the system which they occupy. However, there is hardly any species of tree that does not serve more than one purpose. In a rural economy the basic needs of timber, firewood, fodder, fruit, shade, soil amelioration and overall improvement in the environment are met from the trees. Many trees possess the capability to fix nitrogen through their microbial association.

Within the category of MPTs, the nitrogen fixing trees are those that are endeavored with the capacity to fix atmospheric nitrogen in a form usable by individual plants, enrich the soil through leaching from roots, leaf fall and deliberate mulching and to provide domestic fodder to domestic animals, used as living fences and fuelwood (Burley, 1983). A comprehensive list of 37 species of major importance and 26 of minor importance of NFTs was prepared by Brewbaker and Styles (1982). Another list of more than 1000 species of wood perennials (legumes and non legumes) having potential for nitrogen fixation was prepared by Halliday and Nakao (1982).

Most of NFTs belong to the family Leguminoseae but there are more than 10 other families, known to fix nitrogen (Brewbaker *et al.*, 1984). Promising species of multipurpose trees and shrubs for different agroecological zones of India were identified by Patil and Pathak (1977) and a modified list is presented in annexure 1.

The multiple benefits of MPTs

The trees with many uses would be preferred to those with one single use. But in a broad sense every tree has at least the triple functions of protective, productive and socio-economic roles as follows :

Protective (environmental) :

- moderation of micro/macro climatic parameters, checking soil erosion and water run-off.
- Soil improvement/fertility build up, water conservation and flow moderation.
- Live hedge/boundary demarcation.
- Wildlife habitats.
- Pest and weed control.
- Watershed protection and rehabilitation of degraded lands.

Productive

- *Wood* - timber, building material, veneers, chipboards and other panel products, pulp and paper, rayon etc.
- *Bark* - raw as fuel, dyes, tannins and chemical extraction etc.
- *Energy*

  raw - firewood

  processed- charcoal, gases or liquid fuels, chemical stem extractives, resin, oil, paint, varnishes, pharmaceuticals.
- *Leaf* - Thatch, fodder, oil, silk, bidi wrapper, medicines, honey, dyes, food.

- *Root* - fibre, fuelwood, dyes, chemical extractives.

Socio-economic

- employment generation
- income generation - foreign exchange
- import substitution
- public education
- amenity and tourism
- rehabilitation of degraded lands
- counter seasonality
- risk reduction
- labour saving
- improved human and animal nutrition and health.

In case of MPTs normally, socio-economic gains from its multiple products are the prime considerations. Under specific climatic and edaphic mantle for which the local population selects ideal MPT species for multiple gains. However, adoption of such species in various farming systems requires the attitudinal changes, understanding and cooperation of professional foresters, horticulturist, agronomists and extension workers. MPTs are usually useful for social forestry programmes dependent on peoples needs vis-a-vis National priorities.

Promising MPTs

Based on regional practices and agroclimatic needs, though the species are very specific, yet if we see from the view point of their use as fodder, fuelwood, timber, fruit, shade and medicine some of the species are highly preferable (Table 4.1). Species like *Acacia nilotica, Azadirachta indica, Artocarpus heterophyllus, Emblica officinalis, Dalbergia sissoo, Hardwickia binata, Moringa oleifera, Leucaena leucocephala, Prosopis cineraria, P. juliflora, Pithecelobium dulce, Zizyphus mauritiana* could be considered as highly preferred species besides several others that are usually grown.

Table 4.1: Preferred indigenous and exotic MPTs for use in agroforestry systems

Botanical Name	Habitat Preference		Purpose
Leucaena leucocephala	SA SH	M	Soil amelioration, fodder, fuel biomass for pulp, shade.
Faidherbia albida	SA SH	D	Soil amelioration, fruits as animal feed, fuel, small timber, fencing.
Paulownia tomentosa	Tem. cold Hilly		Soil amelioration, pulp, timber.
Sesbania grandiflora	HT SA	 M	Fruits, flowers for vegetable, leaf as fuel, soil amelioration.
S. sesban	SA HT	M	Soil amelioration, fodder, fuel.
Ailanthus excelsa	SA		Fodder, fuel.
Azadirachta indica	A, SA M	 Trop.	Fodder, fuel, timber, medicine, oil.
Acacia nilotica var. *cupressiformis*	SA Moist	HT	Fuel, timber, fodder, tanins.
Albizia amara	SA	A	Fodder, fuel, small timber.
Prosopis juliflora	HT	SA	Fuel, timber, pods as fodder.
Prosopis cineraria	A		Fuel, fodder, food, medicine,small timber.
Harwickia binata	A	SA	Fodder, fuel, timber.
Gliricidia sepium	HT		Soil amelioration, fodder.
Paraserianthes falcataria	HT		Pulp wood.
Populus deltoides	ST		Pulp wood.
Eucalyptus spp.	HT, ST	SA	Pulp wood, timber, fuel.
Zizyphus mauritiana	A	SA	Fruits, fuel, fodder, timber.
Emblica officinalis	SA	HT	Fruits, fuel.
Moringa oleifera	HT		Fruits, pulpwood, vegetables, medicine.

SA = Semi arid
HT = Humid tropics
A = Arid
ST = Sub Tropical
SH = Sub humid
D = Dry
Temp. = Temperate
M = Moist

An identification of MPTs parameters like water use efficiency, calorific value, density, utility index, etc., the local adaptability in the socio-economic set up and edaphic tolerances determine the preference of a species. In a detailed study involving large number of woody species Nautiyal and Purohit (1988) identified 20 species of which first 5 from superiority index are presented in Table 4.2 in the Central Himalayas. Thus, while a species is superior for one character, for the other it may not be. But for working the overall superiority index all the factors are considered.

Table 4.2 : Superior Multipurpose trees in Central Himalayas (Based on Nautiyal and Purohit 1988)

Sl. No.	Name	WUE	CV	Density (Wood)	Utility index*	Rank**
1.	*Toona ciliata*	11	2	15	4.0	10
2.	*Grewia optiva*	6	5	13	3.5	12
3.	*Ougeinia dalbergioides*	1	7	9	5.5	3
4.	*Bauhnia retusa*	9	8	5	6.0	9
5.	*Dalbergia sissoo*	4	9	6	5.5	1

SI = Superiority index
WUE = Water use efficiency, CV = Calorific value,
*Utility index = Based on major and minor uses
**Rank = over all based on 20 species with 10 characters.

Edaphic considerations in species choice

On normal soils species establishment and growth does not pose any problem but with difficult sites *viz*, acidic, sandy, saline, poorly drained, calcareous, saline sodic and gravelly soils, the species identified from the point of their establishment and growth have given in Table 4.3. It could be amply seen that species like *Prosopis juliflora, Acacia nilotica, Eucalyptus camaldulensis, Zizyphus mauritiana* have highest adaptability and ecological amplitude.

Table 4.3 Multipurpose trees and shrubs suitable for different edaphic situations

1. **Acidic soils :** *Albizia falcataria, A. procera, A. stipulata, Acacia auriculiformis, Alnus nepalensis, Gmelina arborea.*
2. **Sandy arid :** *Prosopis cineraria, P. juliflora, Acacia tortilis, Calligonum polygonicles, Tamarix aphylla, Zizyphus mauritiana, Z. nummularia.*
3. **Coastal sandy :** *Prosopis juliflora, Casurina equisetifolia, Anacardium occidentale.*
4. **Poorly drained :** *Eucalyptus camaldulensis, Albizia procera, Terminalia arjuna, Acacia nilotica, Syzygium cumini, Casuarina equisetifolia.*
5. **Alluvial soils :** *Acacia nilotica, Melia azadirach, Azadirachta indica, Dalbergia sissoo, Eucalyptus tereticornis, Populus deltoides, Leucaena leucocephala, Mangifera indica, Bassia latifolia.*
6. **Calcareous soils :** *Acacia nilotica, Albizia lebbeck, Azadirachta indica, Eucalyptus camaldulensis, Melia azadirach, Tamarix aphylla.*
7. **Saline sodic soils :** *Acacia nilotica, A. tortilis, Pithacelobium dulce, Prosopis juliflora, P. cinereria, Sesbania sesban, S. aculeata, Tamarix aphylla, Azadirachta indica, Albizia procera, Terminalia arjuna.*
8. **Shallow gravelly soils :** *Albizia amara, Zizyphus mauritiana, Hardwickia binata, Anogeissus pendula.*

Use of MPTs as fodder

In India the prominent role expected of the MPTs relates to fodder (during the lean period) and firewood. Many trees have been studied in detail for these traits. A list of trees lopped for fodder in tropical regions is presented in Table 4.4. It can be seen that great variability exists for their use by different species of animals.

Table 4.4 : List of potential trees/shrubs that can be lopped for feeding livestock in tropical parts of India

Species	Preference to livestock							
	Cattle		Goat		Sheep		Camel	
	Leaf	Pod	Leaf	Pod	Leaf	Pod	Leaf	Pod
Acacia nilotica Willd. (Desi Babool)	**	**	***	**	***	**	**	***
Acacia senegal Willd. (Kher)	*	*	**	**	**	**	**	**
Acacia tortilis Hayne. (Israeli Babool)			**	**	**	**		**
Ailanthus excelsa Roxb. (Ardu)		***		***				
Albizia amara Boiv. (Kala Siris)	*		**		**			
Albizia lebbek Benth. (Desi Siris)	**		**		**			
Albizia procera Benth. (Safed Siris)	*		*		*			
Bauhinia variegata Linn. (Kachnar)	*		**		**			
Dalbergia sisso Roxb. (Shisham)	*		**		**			
Ficus glomerata Roxb. (Gular)	**	*	***	**	**	**		
Hardwickia binata Roxb. (Anjan)	**		**		**			
Leucaena leucocephala (Lam.) de Wit (Subabool)	***		***		***			

Table 4.4 - Contd.

Species	Preference to livestock							
	Cattle		Goat		Sheep		Camel	
	Leaf	Pod	Leaf	Pod	Leaf	Pod	Leaf	Pod
Parkinsonia aculeata Linn. (Horsebean)			**	**	**	**		
Pithecellobium dulce Benth. (Jangali Jalebi)		***	**	***	**	**		
Prosopis cinerea Linn. (Khejri)	**	**	***	**	***	**	***	**
Prosopis juliflora DC (Vilayati Babool)		***		***		***	*	***
Sesbania grandiflora Pers (Agastha)	***	**	***	**	***			
Ziziphus nummularia Roxb. (Pala Bordi)		***	**	***	**	**		

*** Excellent

** Good

* Fair

The trees are not preferred as fodder throughout the year. Only during the lean periods they are lopped for meeting the green fodder and nutrient needs. An attempt has been made to identify the availability periods (Table 4.5) which indicates suitability of certain species for fodder throughout the year. Many trees do not provide leaf fodder throughout the year on account of unpalatibility, but their fruits/pods are highly palatable and nutritious. Some species are highly preferred in one season while in the other species their leaves are toxic and not lopped or fed. Similarly, a tee species may be a good fodder in one geographical region, while in other regions it not at all used. A comprehensive detail on fodder trees is presented by Singh (1982).

Table 4.5: Period of fodder availability from some potential trees/ shrubs in tropical parts of India

Species	Period of fodder availability	
	Leaf Fodder	Pod Fodder
Acacia nilotica	May-February	April-June
A. senegal	April-December	February-March
A. tortilis	May-February	May-June
Ailanthus excelsa	May-March	-
Albizia amara	June-March	-
A. lebbeck	April-November	-
A. procera	July-March	-
Bauhnia variegata	April-November	-
Dalbergia sissoo	February-December	-
Dichrostachys cinerea	July-April	August-December
Ficus glomerata	December-August	March-July
Hardwickia binata	May-January	-
Leucaena leucocephala	Throughout year	December and May
Parkinsonia aculeata	Throughout year	December-March
Pithecellobium dulce	Throughout year	April-June
Prosopis cinereria	Throughout year	June-August
P. juliflora	-	April-May or June
Sesbania grandiflora	July-November	-
S. sesban	Throughout year	March-June
Ziziphus nummularia	Throughout year	December January

Table 4.6: Palatability ratings and chemical analysis of freshly lopped fodder of promising fodder trees/shrubs in tropical parts of India

Species	Type of fodder	Palatability ratings	On dry matter basis						
			Crude protein (%)	Crude fibre (%)	Nitrogen free (%)	Ash (%)	Phosphorus (%)	Calcium (%)	Magnesium (%)
A. nilotica	Leaf	Good	13.9	9.2	69.8	7.1	0.1	2.6	0.4
	Pod	Good	13.6	14.9	61.0	6.2	0.2	0.9	-
A. senegal	Leaf	Moderate	10.3	9.5	65.7	16.4	0.05	6.9	0.6
	Pod	Moderate	19.6	39.0	30.9	7.1	2.4	1.1	0.5
A. excelsa	Leaf	Good	19.8	14.3	41.9	19.9	0.3	2.0	-
A. lebbek	Leaf	Moderate	29.2	25.3	43.8	7.5	0.2	1.8	0.5
B. variegata	Leaf	Fair	13.3	29.1	46.3	9.3	0.3	2.9	-
D. sisso	Leaf	Fair	16.6	22.3	49.4	8.5	0.2	2.1	-
D. cinerea	Leaf	Good	15.2	28.1	50.4	4.6	-	-	-
	Pod	Good	11.4	25.6	56.4	5.5	0.2	0.5	-

Table 4.6 - Contd.

Species	Type of fodder	Palatability ratings	On dry matter basis						
			Crude protein (%)	Crude fibre (%)	Nitrogen free (%)	Ash (%)	Phosphorus (%)	Calcium (%)	Magnesium (%)
L. leucocephala	Leaf	Good	21.4	14.2	49.5	4.3	0.2	2.7	-
P. aculeata	Leaf	Moderate	13.1	24.9	40.9	11.9	0.2	2.9	-
P. cinereria	Leaf	Good	13.9	20.3	59.2	6.5	0.2	1.5	0.5
P. juliflora	Leaf	Fair	21.4	20.8	50.0	7.7	0.2	1.5	0.5
	Pod	Good	20.8	30.8	35.0	-	0.2	0.3	-
S. grandiflora	Leaf	Good	26.2	8.8	51.4	10.2	-	-	-
Z. nummularia	Leaf	Good	11.7	16.0	65.0	7.2	0.2	1.6	0.3

Palatability and nutritive value has been worked out for most of the tree species. Many species possess more than 20 per cent crude protein and are highly palatable (Table 4.6).

Though lopping is the principal mode of collecting nutritious leaf fodder, it reduces the growth increment in many lopped tree species unless done up to optimum level (Roy 1990, Table 4.7). Out of six species studied, *Albizia amara* was found to be affected maximum while *A. procera* was the least affected.

Table 4.7 : Trend of tree growth in six fodder tree/shrub species in lopped and unlopped treatments

Species	Treatment	Average initial growth (cm)		5-year mean annual increment (cm)	
		CD	DBH	CD	DBH
A. tortilis	Unlopped	22.2	17.3	2.0	1.64
	Lopped	21.6	17.2	1.9	1.58
A. amara	Unlopped	36.8	70.3	1.9	15.50
	Lopped	35.1	89.0	1.2	9.30
H. binata	Unlopped	21.6	16.9	1.7	1.50
	Lopped	20.9	16.7	1.6	1.40
A. lebbek	Unlopped	24.6	18.9	1.4	1.30
	Lopped	23.9	18.3	1.3	1.20
A. procera	Unlopped	24.9	23.4	1.9	1.70
	Lopped	25.2	24.7	1.9	1.60
D. cinerea	Unlopped	13.6	13.6	1.1	1.60
	Lopped	12.9	7.9	1.0	0.90

Biomass production potential

Increasing demand for woody biomass to support the industry for pulp, small timber and firewood can only be met through fast growing

Table 4.8 : Yield of MPTs in energy plantations

Species	Rotation (Years)	Density no./ha	Production Air dry t/ha/yr		
Eucalyptus tereticornis	6	964	7.9	(1)	
	6	1038	5.1	(2)	On different site quality
	7	1105	3.0	(3)	classes from 1-5.
	8	1202	1.6	(4)	
	9	1345	0.7	(5)	
Casuarina equisetifolia	5-15	2500	7-20		
Prosopis juliflora	15	-	5-7		
	30	-	2.4 - 3.2	(Semi arid areas)	
	10	-	5	(500 mm rainfall)	
Acacia nilotica	-	-	5.8 - 7.6	(1)	Values in
	-	-	3.4 - 4.8	(2)	paranthesis are site
	-	-	1.4 - 2.4	(3)	quality
Eucalyptus tereticornis	21(m)	-	13.6	(ADT/ha)	
Albizia lebbeck	18(m)	-	12.7	"	
Dalbergia sissoo	18(m)	-	14.9	"	
Acacia nilotica	18(m)	-	30.0	"	
Cassia siamea	18(m)	-	18.6	"	
Acacia tortilis	18(m)	-	17.7	"	
Prosopis juliflora	18(m)	-	19.7	"	

MPTs grown on private lands or the common wastelands. Growing these trees in farm forestry/agroforestry coppice farms under energy plantation holds great promise for this need. It has been observed that species like *Acacia nilotica* could produce 30 t/ha at 18 months (Table 4.8). In another study Pathak (1987) reported 74.3 t/ha/Yr biomass productivity in *Leucaena leucocephala* at 5 year rotation. Even at 1 or 2 year rotation *L. leucocephala* yielded 19.2 to 48.9 t/ha/Yr on moist habitats (Pathak and Gupta 1987). Many such species are highly adaptable to high planting density and possess wide ecological amplitude, thus producing very high biomass in time and space.

Most of these multipurpose trees and shrubs produce wood of high calorific value and good burning quality. A list of few MPTs with their firewood quality is presented in Table 4.9.

Table 4.9 : Quality of firewood of sundried lopped branches of promising fodder trees/shrubs in tropical parts of India

Species	Fuelwood quality	Charcoal quality	Calorific value (K cal per kg)	Other specific use
A. nilotica	Good	Good	4875	-
A. senegal	Moderate	Moderate	-	-
A. tortilis	Good	Good	4400	Fencing
A. excelsa	Fair	-	-	-
A. amara	Good	Fair	5600	-
A. lebbeck	Good	Good	5200	-
A. procera	Fair	Fair	5500	-
D. sisso	Excellent	Good	5050	-
L. leucocephala	Good	Good	4400	Pulp
P. aculeata	Good	Good	-	-
P. dulce	Good	-	5400	-
P. cinerea	Excellent	Excellent	5000	-
P. juliflora	Good	Good	5000	Fencing
S. grandiflora	Poor	-	-	Pulp

Cultivation technology

Apparently it appears that every think about the cultivation of a multipurpose trees are known. But the changing situation of soil nutrients, their physical characters and also the socio-economic structure often makes it impossible to establish a species and even if established, to grow them in desirable forms and make them produce at the optimum level. Some basic issues regarding their cultivation are as under :

- objectives of planting and needs to be met
- identification of the proper species
- search for improved seeds/planting materials
- land availability
- proper nursery techniques/inputs
- pre planting preparation
- planting techniques and after care
- protection from biotic agencies and management
- utilization/marketing/economic returns

The most important single factor in this programme is the land availability. The state of common property resources is well known and the per capita availability of the land area for agriculture is also very low. Yet the need for trees and their products is felt and the trees are planted around fields, along canals, around the homesteads and also in the fields with the crops.

Planting objectives

While making survey for diagnosis and design exercise around Jhansi, it was observed that the single driving force for any activity was income/money flow from the system for which the farmers prefers to plant fruit trees. The urgent needs/priorities could be summarized under food, cash, fuelwood, timber and security. Based on the land based activities it was observed that trees were able to meet their needs for cash, fuelwood, fodder, fruit and timber.

To meet these demands farmers plat woody species wherever they can manage and protect. Thus to meet these objectives multipurpose trees preferred are :

- fruit cum shade giving species
- fodder cum fuelwood species
- timber cum fodder species

Choice of species

Species choice is based on the adaptability to the site, early returns, multiple uses, complimentarity to the system and its possible role during the lean/critical periods. Characteristics that make a tree ideal for agroforestry systems were delineated by King as early as 1979. These are as under :

- amenability to early wide spacing
- self pruning habit
- tolerance to high level of pruning
- favourable low crown diameter to bole diameter ratio
- light branching
- tolerant of side shade in early growth stage
- appropriate phyllotaxis that permits the penetration of light to ground
- the phenology, particularly with respect to leaf flushing and leaf fall should be advantageous to the growth of annual crops with which they are grown
- the rate of litter fall and its decomposition should have positive effect on the soil
- their above ground changes over time in structure and morphology should be such that they retain or improve those characteristics which produce competition for solar energy, nutrients and water
- the root system and root growth characteristics should result in exploration of soil layers that are different to those being tapped by the agricultural species, and
- they should be efficient nutrient pumps

Similarly for making ideal fodder tree species following characteristics are considered important :

- high palatability with good nutritive value, minimum anti quality factors/substances in the leaves and young shoots
- high coppicing ability for several sustained cycles
- a multilayered canopy to provide high energy synthesis efficiency

- hardiness to drought and frost
- tolerance to occasional flooding
- indeterminate growth

Often selection is made on the basis of returns without considering the local factors of soil, climate and the market for the produce. While preference requires to be given to the indigenous species, some well tested exotics need incorporation in the system. A list of such species is presented in Table 4.10.

Table 4.10 : Promising multipurpose trees and shrubs

Botanical name	Use					
	F_o	F_u	T	F_r	S	M
Acacia nilotica	3	5	4	-	2	3
A. tortilis	1	4	2	-	1	-
Albizia lebbeck	2	3	3	-	3	-
A. amara	3	4	2	-	3	-
A. procera	2	3	3	-	2	-
A. falcataria	-	4	3	-	2	-
Ailanthus excelsa	4	1	1	-	3	-
Azadirachta indica	4	4	4	-	5	5
Anona squamosa	3	-	-	4	-	-
Artocarpus heterophylla	2	3	4	5	4	-
Cassia siamea	-	4	3	-	5	-
Casuarina equisetifolia	-	5	3	-	-	-
Ceiba pentendra	1	1	2	5	2	-
Derris indica	3	3	2	-	4	-
Dalbergia sissoo	2	4	5	-	3	-
Dendrocalamus strictus	3	2	3	-	-	-
Dichrostachys cinerea	3	4	-	-	-	-
Emblica officinalis	2	3	3	5	2	-
Eucalyptus tereticarnis	-	3	4	-	-	-
Gliricidia sepium	2	2	-	-	2	-

Botanical name	Use					
	F_o	F_u	T	F_r	S	M
Hardwickia binata	4	3	5	-	2	-
Melia azadirach	-	4	3	-	2	-
Moringa oleifera	4	2	1	5	1	-
Sesbania grandiflora	5	2	-	4	1	-
S. sesban	5	2	-	-	1	-
Leucaena leucocephala	5	4	4	-	3	-
Zizyphus mauritiana	2	4	-	4	-	-
Prosopis juliflora	-	5	2	3	2	-
P. cinereria	5	5	4	3	3	-
Pithecellobium dulce	4	4	3	2	2	-
Robinia pseudoacacia	2	4	3	-	2	-

Grades 1-5 indicate increasing scale of use from small to higher value.

Search for improved seeds/planting materials

It is essential to search for proper seed/planting material (variety/ cultivar) and its authentic source before initiating activity related to MPTs plantations. In case of fruit trees grafted/buded materials with improved varieties are preferred to seedlings. In case of several tree species much is not known except the plus trees selected by the forest Departments. Tree improvement programme in India is still is infancy. However, improved varieties/cultivars are available in case exotic poplars and *Leucaena leucocephala*. Table 4.11 presents comparative performance data in 3 plant types of *L. leucocephala*. The difference in biomass production between the Hawaiian and Silvi-4 variety is striking. Silva-4 is even much superior to K-500 variety. It is now very essential that proper seed material from selected varieties is used for large scale plantation for specefic purposes.

Proper nursery techniques/inputs

Raising nursery to provide suitable seedlings for planting in the fields is a very careful process. Well trained staff with optimum facilities can only handle this exercise. Roy and Pathak (1990) outlined the guidelines for seed collection of some fodder trees and also the presowing treatments to the seeds to obtain optimum germination. Seed storage methods assure

Table 4.11 : Comparative performance of 3 varieties of* Leucaena *under farm forestry

Attributes	Varieties		
	Hawaiian	K-500	Silvi-4
Collar diameter (cm)	17.5 ± 0.9	19.2 ± 0.8	28.6 ± 6.7
Diameter at breast height (cm)	15.7 ± 3.8	16.8 ± 0.3	25.4 ± 5.1
Height (m)	13.2± 1.3	11.1 ± 0.8	19.9 ± 1.5
Production (kg/tree)			
bole	52.7 ± 9.8	59.8 ± 19.5	418.4±152.5
branch	28.1 ± 7.2	23.4 ± 2.9	113.2 ± 44.1
Leaf	2.1 ± 0.5	1.7 ± 0.7	26.7 ± 7.2
pod	0.3 ± 0.1	0.6 ± 0.1	1.7 ± 1.4

proper germination since many species require after ripening time (*Prosopis juliflora, Albizia lebbek* etc.). The hard seed coat in many species requires scarification by mechanical, chemical or hot water methods to give optemum germination. Table 4.12 gives details of some methods for successful raising of nursery in MPTs. The soil mixture in polythene bags and their sizes have also been determined for many species. In most of the MPTs/NFTs, mixture of sandy soil + Farm yard manure or red soil + black soil in 4:1 ratio have given the best root growth and nodulation. Heavy black soils are not suitable for raising the seedlings. Black polythene bags are considered superior to normal transparent ones. In case of species where nursery is to be maintained for 2-3 months, 10 x 20 cm, 150 gauge polythene bags are sufficient, but where 6-12 months old seedlings are desired 15 x 25 cm, 200 gauge polythene bags are necessary. Though the nurseries are to be made at a place where regular frequent irrigation, good soil, FYM, insecticides, and other inputs are readily available. Partial shading during summers is essential. For this Typha grass or Palm leaves are very useful. In species like *Leucaena leucocephala*, partial shading enhances seedling growth in the nursery and assures optimum survival after planting in the field (Pathak *et. al.*, 1983). For many other species such specific requirements are still to be worked out.

Table 4.12 : Nursery techniques for some MPTs species

Species	Average no. of seed/kg	Seed pretreatment	Sowing method
A. nilotica	8500	Treatment with Con. H_2SO_4 for 30 minutes followed by overnight soaking	Direct sowing 1-2 seeds per container
A. tortilis	9000	With Con. H_2SO_4 for 25-30 minutes followed by overnight soaking	Direct sowing 1-2 seed per container
A. farnesiana	16400	Scarification by Conc. Sulphuric acid for 5-6 minutes followed by soaking	Direct seeding in polypots
A. excelsa	2700	Presoaking the pods for 3 days before sowing	Pricking out the young seedlings from the bed
A. lebbeck	12000	Presoaking the seeds for 24 hrs before sowing or Hot water 80°C treatment for 30 seconds	Direct sowing 1-2 seed per kg or Pricking out

Table 4.12 : Contd.

Species	Average no. of seed/kg	Seed pretreatment	Sowing method
A. procera	21000	Hot Water 80°C treatment for 30 seconds	Pricking out " " "
A. amara	25000		" " "
D. cineria	42000	Treatment with Con. Sulphuric acid for 15 minutes followed by overnight soaking	Direct sowing 1-2 seed per container
L. leucocephala	20000	Hot water treatment for 30 second or Treatment with con. Sulphuric acid for 5 minutes	Direct sowing 1-2 seed per container
P. cineraria	26000	Feed the pods to sheep and goat and collect the pens.	Direct sowing 1-2 seed per container
P. juliflora	32000	Feed the pods to sheep and goat and collect the pens or Treat the pods with 40 per cent KOH for 2 days collect the seed and then treat the seed with hot water (85°C) for 5 minutes and soak in water overnight	Direct sowing 1-2 seed per container

Pre-planting preparations

Digging pits in winters and allowing these to get seasoned during summer ensures ease in planting and handling large scale planting during July when the rains start. Planting soon after monsoon also assists in higher establishment. On very poor soils, fully decomposed farm yard manure is also required at least 2-5 kg/pit. On slopy lands contour trenches (3 m x 0.5 m x 0.3 m) and bunds are to be prepared before pit digging so as to allow more moisture conservation and checking run-off and soil loss. Just after pre-monsoon showers, seedlings should be transported to the planting site. Wherever plantation is scheduled at a place near habitation, protection becomes essential to save it from biotic agencies which should preferably done through social fencing.

Planting techniques and after care

Out of several methods, the most suitable is planting the polypot raised seedlings of good vigour and appropriate age. In certain large scale programmes, good success has been achieved by seed sowing for *Acacia nilotica*, *Prosopis juliflora*, *Dichrostachys cinerea*, *Azadirachta indica*, *Butea monosperma* etc. during monsoon after appropriate soil preparations. However, seedling planting has an edge over the sowing method. During planting after half filling the pre-dug pits, a small dose of 100 g fertilizer mixture (urea: single super phosphate and Potash in ratio of 1:4:1) as basal application followed by planting has given higher establishment and early good growth in many species on very poor soils (Gupta *et. al.*, 1988). The small additional expenditure is amply compensated by faster growth of seedlings compared to unfertilized ones. It is very important that when polythene is removed before planting, the root coiling, if any should be removed by cutting it out, otherwise the plants do not grow properly in desired form. Normally planting is done during the rainy season, but in case of periodic long drought, irrigation is essential for the success. In the first year of growth during dry hot months, irrigation, where possible ensures higher survival and growth, where irrigation facilities are not available appropriate mulching technology should be adapted.

Protection from biotic agencies

Early growth stages of MPTs are very tender and susceptible to damage by termites and rabbits. Addition of 25 g/pit B.H.C. or 5 g/pit Aldrin dust ensures freedom against termite attack. Wild rabbits create problems in case of highly palatable species like *Leucaena leucocephala* and *Sesbania sesban*. BHC dusting repells rabits. Grazing animals both domestic and wild are yet another threat to their establishment and growth. Fencing by barbed wire or chain link is very costly. These days cattle proof trenches are being provided to keep off the animals. On the berms of cattle proof trenches, sowing of thorny tree seeds viz., *Prosopis juliflora, Acacia nilotica* and *Dichrostachys cineria* in three lines 50 cm apart makes a green thorny fence which also provides additional biomass resources for use. When such plantations are being planned along farm boundary, farmers normally hesitate due to problems of saving these plants during off season periods when crops are harvested. Due to only this reason they go for unpalatable species. In case wherever they plant highly palatable species individual trees are protected by fencing around them through thorns or other leafy biomass like Palm leaf etc. However, social fencing is the cheapest and most effective methods for protection against grazing and vandalism.

Management of MPTs have to be viewed in the light of objectives with which they are planted. Where they are planted for only fodder, cutting management has to be strictly followed to harvest maximum (60 cm height, 40-50 days duration in case of *Leucaena leucocephala* where plant density is over 60,000/ha). In alley cropping, the height and cutting frequency is adjusted to the needs of crop growth and seasonal requirements. In such a system 4-5 cuttings of *Leucaena leucocephala* are taken for fodder (6-8 t/ha) and in early 3 years harvesting from the base yields firewood (8-10 t/ha). Along farm boundary/road side single rows of trees periodic pruning avoids shade from lower branches in addition to fodder from leaves and firewood from the twigs. Rotational harvest (3-4 years) and prunings of the coppice shoots has produced upto 24 t/row km firewood and 3-6 t/row km fodder every 3-4 years (Pathak 1986). Wherever they are managed as compact block of energy plantations, Pathak (1987) reported, in case of *Leucaena leucocephala*, a boimass production up to 74.3 t/ha/year. Management of compact blocks for energy plantation in case of minirotation species has given very encouraging results in many other species.

Utilization, marketing and economic returns

Multipurpose trees are intended to help rural population to meet their needs of small timber, firewood, fodder, fruit, green manure, brush wood for fencing, shade and overall environmental amelioration. In case of excess it should provide grower with cash flow through sale in local markets. Sale of tree leaf fodder or firewood in the market is not rare but the grower hardly gets remunerative price for it. The in-situ use as firewood is expected to help soil amelioration through availability of increased farm yard manure for field crops. This resource has a great promise in sustaining profitable agriculture in the country.

Economic viability of MPTs has been studied in detail (Abrol and Joshi 1984, Relwani and Hegde 1986, Shah 1984, Pathak and Khan 1989). In our studies under minirotation energy plantations of subabul, it was observed that the benefit: cost ratio ranged from 0.69, 4.72 to 7.8 with 4000, 5000 and 10,000 plants/ha density at 2, 5 and 5.5 years respectively. Very high rate of return has been observed when plants are grown on fertile soils.

Suggested future research thrusts are:

- Identification of most promising germplasm/variety of trees and their improvement for higher returns
- Seed multiplication of desired varieties and authentic seed supply for plantation
- Nursery techniques for plantation on refractory sites and problem soils
- Peoples "education" to take up tree farming as an important activity [forestry extension research]
- Standardization of management and utilization practices to sustain productivity of the tree and higher returns to the farmers
- Creation of ideal marketing infrastructures to ensure high returns to growers
- Creation of industrial infrastructure to process plant products into valuable material wherever feasible
- Integration of MPTs in farming systems to sustain the edaphic environment and assure higher returns.

Acknowledgements

Grateful thanks are due to Dr. Panjab Singh, Director, Indian Grassland and Fodder Research Institute, Jhansi for the encouragements and facilities to carry out the research.

REFERENCES

Abrol, I.P. and Joshi, P.K. (1984). Economic viability of reclamation of alkali lands with special reference to agriculture and forestry. In: (Kamal Sharma ed.) *Economics wastelands development*; SPWD; pp. 19-30.

Blair, G., Catchpole, D. and Horne, P. (1990). Forage tree legumes: their management and contribution to the nitrogen economy of wet and humid tropical environments. *Adv. Agron.* 44: pp. 27-50.

Brewbaker, J.L. and Styles, B. (1982). Economically important nitrogen fixing tree species. NFTA miscellaneous publication, pp. 82-104, NFTA, Waimanalo, Hawaii.

Burley, J. (1983). Global needs and problems of collection, storage and distribution of multipurpose tree germplasm. In. *Multipurpose tree germplasm.* Proceedings of a planning workshop to discuss International cooperation. pp. 29-30, Nairobi, Kenya, ICRAF.

Brewbaker, J.L., Van Den Beldt, R.J. and Mac Dicken, K. (1984). Fuelwood uses and properties of nitrogen fixing trees. *Pesquisa agropec bras* 19, pp. 193-204.

Felker, P. and Bandursky, R.S. (1979). *Econ. Bot.* 33(2), pp. 172-184

Gupta, S.K., Pathak, P.S. and Khan, T.A. 1988. Fertilization application boosts early growth of *Acacia tortilis* Forsk. and *Dalbergia sissoo* Roxb. on degraded rangelands in central India. *Third International Rangeland Congress* Abstracts. II. pp. 556-559.

Halliday, J. and Nakao, P. (1982). The symbiotic affinities of woody species under consideration as nitrogen fixing tree. Univ. of Hawaii, NiFTAL Project, Honolulu, Hawaii.

King, K.F.S. (1979). Key note address—Some principles of agroforestry. Proc. Agroforestry Seminar, pp. 17-26, ICAR, New Delhi.

MacDicken, K.G. and Sastry, C.B. (1988). Preface. In. *Multipurpose tree species for small Farm Use*, p. 7. Winrock International/IDRC.

Nautiyal, A.R. and Purohit, A.N. (1988). Superiority indices of some multipurpose trees from the central Himalayas. In. (Withington, D., K.G. MacDicken, C.B. Sastry and N.R. Adams eds.) *Multipurpose tree species for small farm use.* pp. 254-260, Winrock International/ IDRC.

Pathak, P.S. Rai, M.P. and Deb Roy, R. (1983). Seedling growth of *Leucaena leucocephala* (lam.) de wit. I. Effect of shading. *Indian J. For.* 6(1), 28-31.

Pathak, P.S. (1986). Fuel and forage production from *Leucaena leucocephala* under farm forestry. *Indian For.* 112(6), pp. 485-490.

Pathak, P.S. (1987). *Leucaena leucocephala* based production systems for wastelands in social forestry. In.: (Khosla, P.K. and R.K. Kohli eds.) *Social Forestry for rural development*, pp. 86-95.

Pathak, P.S. and Khan, T.A. (1989) Economics of minirotation energy plantations on moist waste lands. Proc. Economics of energy plantations on the waste lands, Udaipur.

Patil, B.D. and Pathak, P.S. (1977). Energy plantation and silvipastoral systems for rural areas. *Invention Intelligence* 12, pp. 79-87.

Relwani, L.L. and Hegde N.G. (1986). Economics of subabul plantation. In. (Hegde, N.G. and P.D. Ablyankar ed.). *The greening of wastelands*. BAIF, pp. 118-121.

Roy, M.M. and Pathak, P.S. (1990). A guide to seed collection of some tropical fodder trees. *My forest* 26(1): pp. 63-69.

Roy, M.M. (1990). Management of fodder trees for optimum fodder production. Workshop on MPTS for rural needs. IGFRI, Jhansi.

Shah, P. (1984). Economics of wasteland development projects undertaken by GSRDC. In. (Kamal Sharma ed.). *Economics of wastelands development*. SPWD, pp. 69-82.

Singh, R.V. (1982). *Fodder trees of India*, p. 663, Oxford and IBH, New Delhi.

Van Carlowitz, P.G. (1984). Multipurpose trees and shrubs: opportunities and limitations. The establishment of a multipurpose tree data base. ICRAF working paper no. 17, Nairobi, Kenya.

Van Carlowitz, P.G. and Burley, J. 1984. Multipurpose tree germplasm, Nairobi, Kenya, ICRAF.

Annexure I

Multipurpose trees and shrubs for different agroclimatic regions in India

1. Temperate and humid western Himalayan region :

   Populus ciliata, Populus euphratica, Quercus dilatata, Q. incana, Q. semi-carpifolia, Salix alba, Robinia preudoacacia, Morus nigra, Celtis australis, Grewia oppositifolia, Rubus sp., Grevillea robusta.

2. Humid Bengal - Assam region :

   Azadirachta indica, Dalbergia sissoo, Emblica officinalis, Pithecellobium dulce, Pongamia pinnata, Albizia lebbeck, A. stripulata, A. f alcataria, Artocarpus heteropyllus, Bauhinea sp., *Leucaena leucocephala, Movinga oleifera, Morus alba, Syzygium cumini, Eucalyptus tereticornis, Leucaena leucocephala, Sesbania grandiflora, S.sesban, Acacia auriculiformis, A. mangium, Ailanthus excelsa, Alnus nepalensis, Enterolobium cyclocarpum, Gliricidia sepium.*

3. Humid eastern Himalayan region and Islands :

   Eucalyptus globulus, Pongamia pinnata, Quercus semicarpifolia, Albizia chinensis, Artocarpus heterophyllus, Moringa oleifere, Morus serrata, Alnus nepalensis, Syzygium agueum, Pongamia pinnata, Mesua ferrea, Leucaena leucocephala, Sesbania sesban, S. grandiflora.

4. Sub-humid Sutlej-Ganga alluvia plains :

   Acacia nilotica, Anogeissus latifolea, A.pendula, Azadirachta indica, Dalbergia sissoo, Kydia calycina, Morus alba, Pithecellobium dulce, Pongamia pinnata, Zizyphus mauritiana, Ailanthus excelsa, Albizia lebbeck, A. procera, A. amara, Bauhinea variegata, Cordia dichotoma, Moringa oleifera, Adina cordifolia, Bridelia retusa, Schelichera triquga, Gmelina arborea, Leucaena leucocephala, Sesbania grandiflora, S. sesban, Dichrostachys cineria, Desmodium cephalotus.

5. Sub-humid/humid eastern and south eastern upland region :

   Acacia nilotica, A. auriculiformis, Azadirachta indica, Cordea dichotoma, Dalbergia sissoo, Delonix alata, Emblica officinalis, Grevillea robusta, Hardwickia binata, Kydia calycina, Pithecellobium dulce, Pongamia pinnata, Zizyphus mauritiana, Ailanthus excelsa, Albizia lebbeck, A. amara, Artocarpus heterophyllus, Bauhinea purpurea, B.variegata, Moringa oleifera, Morus alba, Leucaena leucocephala, Glyricidia sepium.

6. Arid western plains :

   Acacia tortilis, A. senegal, Azadirachta indica, Cordia rothi, Hardwickia binata, Morus alba, Pithecellobium dulce, Salvadora oleoides, Tamaxix sp., *Zizyphus mauritiana, Ailanthus excelsa, Tecomella undulata, Eucalyptus camaldulensis, E. tereticornis, Prosopis cineraria, P. juliflora, Dichrostachys cineria, Colophospermum mopane.*

7. Semi-arid lava plateau and central highlands :

   Acacia nilotica, A.mearnsii, Casuarina equisetifolia, Emblica officinalis, Kydia calycina, Pithecellobium dulce, Pongamia pinnata, Tamirindus indica, Zizyphus racemosa, Hardwickia binata, Moringa oleifera, Eucalyptus sp. *Glyricidia sepium, Thespesia populnea, Dichrostachys cinerea, Leucaena leucocephala, Sesbania grandoflora, S. sesban, Melia azadirach.*

8. Humid and semi-arid western ghats :

   Acacia auriculiformis, A. melanoxylon, A. nilotica, A. planifrons, A. nilotica, Var. *cupressiformis, Albizia amara, A. procera, Cassisa auriculata, Casuarina equisetifolia, Eucalyptus* sp., *Kydia calycina, Pithecellobium dulce, Zizyphus mauritiana, Aegle marmelos, Ailanthus excelsa, Cordia dichofoma, Hardwickia binata, Leucaena leucocephala, Tamarindus indica, Glyricidia sepium, Thespesia populerea, Sesbania sesban, S. grandiflora, Calliandra calorthyrsus, Cassia siamca, Erythrina suberosa, Melia azadirach, Pterocarpus indicus, Samanea saman.*

Chapter 5

Agroforestry as an Alternate Source of Energy Requirements of our Nation

*A.N. Yellappa Reddy**

1. Energy situation in our Nation

Every development process depends upon suitable inputs of Energy. So sufficient energy supplies are one of the central problems of human development and rank equal to other basic needs, such as food and housing.

The provision of adequate energies is an essential precondition for rural development. Adequate natural resources are grossly under-developed, some inputs to create natural resources not withstanding. Therefore, suitable strategies are necessary not only from the economic point of view, but also from the ecological aspects to create renewable energy resources.

If we look to the energy situation of our country very little commercial energy is used in rural sector. The traditional non-commercial energy sources like firewood, animal wastes and vegetable wastes from Agriculture, form the major source of domestic energy even today.

Over 600 million people in our country cook their food with wood every day. Another natural fuel is cowdung. Burning and mineralising this important organic substance, deprives the soil of vital ingredient to restore the soil fertility. Taken together, the scripping of tree and shrubs from the land for timber and firewood and burning of cowdung as fuel, coupled with overgrazing in forest and common lands are the precise recipe for disastrous desertification that has overtaken our country covering an area of nearly 90 million, hectares which have become a wasted land. This comprises over 25 per cent of the geographical area.

**Research and utilisation circle, Forest Deptt., Bangalore-560003.*

From global point of view, the percentage of arable land to total land area is 10.7 per cent, whereas in India this percentage is 51.5 per cent.

Our moderate climatic and fertile soil with appropriate technology makes it possible to grow two crops and also to produce considerable quantity of biomass required for energy, fodder etc. Paradoxically the situation in rural areas is very acute, so far as the energy is concerned. One member of the family has to devote nearly 150-200 days of the year to procure sufficient firewood for one family. From this scare commodity only 5 to 10 per cent of the energy is utilised in open fire placed (chulahas). Improved ovens can save the wastage of energy for 80-90 per cent. Somehow this has not drawn the attention of the people who plan and government programmes to provide improved ovens at a subsidised rate are too inadequate though it is highly rewarding from the ecological point of view. Many other rural based industries like brick, tile, tobacco curing etc., also contributing greatly in depleting the greenery in rural landscape.

To meet the rural and urban demand many fruit yielding trees have become victims of the rural and urban developmental activities depleting vital food resources and loosing the regular income yielding assets to the rural folk.

In our villages, energy production is closely integrated in a food, fuel and animal production system which interacts with its ecological and economic environment. The working hours spent particularly by women on collecting fuelwood cannot be utilised for work in the field and house which is very important in a rural family. Animal wastes can be employed, as a fertilizer and fuel by adopting simple technology. The usufructs yielding trees are cut for immediate gains resulting in loss of regular income and plant residues that contribute to meet biomass needs like fuelwood, leaf litter and other by products.

In our country, men and livestock are primarily responsible for desertification, soil erosion and water run-off, depletion of underground water resource and edible plant resource etc. Therefore, the major focal point for economic development is to work out programme to provide food, energy and fodder to rural inhabitants.

Further, investments on unproductive and ecologically unfriendly items and concentration of industrial growth around urban conglomerations have concerned nearly 80 per cent of all the technological benefits and resources generated thereon, by 20 per cent of the Urban population. But in rural areas due to low investment and lack of appropriate technology and thrusting of unsuitable, ecologically harmful, highly priced of energy intensive items have increased the cost of cultivation and adversely affected the soil and other natural resources. Inadequate investments on sustainable growth has increased the poverty and strained natural resources. In many cases it has out reached the tolerance limit and is heading for ecological catastrophe. To set-right the man made maladies, heavy capital investment is required for a longer period.

Energy saving strategies

Not much attention has been paid for energy saving devices and to provide ecologically friendly cost effective energy producing technologies in the developmental programmes.

Important energy saving devices are :-

1. Increasing efficacies for water and fertilizer application.
2. Improved maintenance of machinery.
3. Labour intensive weeding of fields instead of applying herbicides,
4. Use of Biological pest control methods and integrated pest management practices.
5. Chemical fertilizer can be replaced by, Nitrogen fixing trees which provides required Organic fertilizer, fodder, fuel, protein rich cattle feed, carbohydrates and with these products, commercial energy can also be produced.

By adopting the above, the energy input required per hectare could be considerably reduced making Agriculture a self regulating economic system. Whereas to-day the farmer has to part with 40 per cent of his income for procuring high non-renewable energy consuming products like chemical fertilizer pesticides, herbicides etc.

Linking Food, Energy and Fodder

Our country is very rich in plant species which provide not only food, fuel, medicine, fibre, but also provide varieties of raw materials for industrial purposes. Processing of the materials produced in rural areas opens up badly needed employment generating opportunities and in generating varieties of end products including GASHOL a product of Carbohydrate fermentation for energy source.

Biomass production in our country is 3 to 4 times higher than the temperate and Savanna zones of the world. Yet our country is suffering from scarcities of essential commodities including the renewable resources like fuelwood due to improper soil and water management, and lack of appropriate technology to utilise wood sources to produce finished products at the rural sites.

Nearly 70-75 per cent of the rural energy is obtained from biomass alone. Unfortunately right from the First Plan no attention was given to replenish the basic resources like fuel and fodder probably presuming that they are inexhaustible. Biomass is indeed nothing more than a Solar Energy which is stored by means of Photosynthesis and which can be converted into consumable energy in various ways :-

(a) Firewood

(b) Charcoal

(c) Straw

(d) Biogas from animal and vegetable waste (Methane gas)

(e) By fermentation (Ethanol/Gashol).

(f) By pyrolysis (wood gas).

(g) Hydrocarbons (from Latex of Euphorbia species).

(h) Non-Edible oils (from Karanja, Neem, Mango, Kernal, Vateria fruit, etc.).

Nearly 30-35 per cent of the land in our country is classified as Arid and Semi-arid. Inadequate, un-reliable and variable rainfall and degraded soils due to over grazing, destruction of vegetation, have converted these lands unsuitable for dependable and meaningful sustainable agriculture. Added to these, regular draughts which is becoming an unsurmountable recurring issues in our Nation contribute for scarcity of fodder and even drinking water. Heavy expenditure on draught relief and allied works are incurred as an annual feature but lasting assets are not created. On the contrary the resources are drastically depleting in rural areas. Substantial environmental degradation has taken place since two decades and nearly 15-20 per cent of the landed area has reached terminal stage.

Poly and Perma-culture

The only alternative is to put these areas under poly-culture and Perma-culture by developing the site specific technology and by planting site specific species. Although the productive capacity of the site appears to be doubtful, with proper technology and inputs and negating the negative aspects, majority of the land can be made productive. The farmers of *Saurashtra* has shown the way to make marginal land to be highly productive by adopting appropriate technology. As stated earlier, nitrogen fixing trees play an important role in enriching the soil in such degraded areas with the help of rhizobia. Most of the Leguminous species have the extra ordinary ability to withstand hostile site conditions and to enrich the soil by producing enormous quantity of leaf litter which is rich in essential nutrients like N.P. and K. These have the ability to restore the Micro-biological activity and also to withstand acute draught etc. All such areas can be put under high density energy plantations incorporating other native tree species depending upon the site capability and biomass need of the locality.

Rural Energy supplies through Biogas and Wood Gasifier

Biogas has not been used extensively for rural electrification. Wood Gassifier by *Pyrolysis* and Methane by *Biogas* can adequately meet the rural requirement at a much lower cost and also, supplies will be assured because the management is within their control. Though above technology is developed, unfortunately it has not yet become popular mainly due to lack of proper support and encouragement from the Government. By popularising and setting right some deficiencies rural areas can became self sufficient in their energy requirements obtained from non-conventional sources.

Future strategy

At present greater use of renewable energy sources is not so much a technological problem. The Scientists have developed the technology which with little refinement, at the time of implementation can mitigate the situation. To implement it is imperative to have adequate capital investment and political will. Simultaneously energy strategies at regional levels has to be developed keeping in view the ecological fundamentals.

As explained above nitrogen fixing trees offer a good prospects of raising alley or companion cropping in agroforestry that can providing the energy in the form of fertilizer, fodder, fuel and food.

Renewable energy is the only alternative

Industrial Society in the World is slowly crippling due to oil crisis. As is known, earth receives vast energy from geographical and photochemical processes. Most of the energy comes from the sun. The sun's energy can be stored only to a limited extent in the form of heat. But there is a process by which some of the energy received from sun can be retained. This process is the basis for all higher forms of life on our planet. The ability of green plants to synthesis, organic compounds from inorganic matter with the aid of sun light, thus fixing solar energy by converting it, as cellulose and carbohydrates etc., is the only source of food for all living beings in the planet.

Earth's fundamental source of energy is available on a unlimited scale particularly in tropical world where we have over 2-thousand light hours. If plants are harvested or decayed, the process of fixation is repeated by new plants and these goes on uninterruptedly. Unfortunately these process has been impaired by man made negative manifestations due to un-scientific management of rich natural resources.

The traditional agricultural practices could be able to retain a small portion of solar energy and the modern agriculture has failed to improve their ability to derive maximum nutrients from the soil and the high technology dragged them to depend more on chemical fertilizers. These unholy external practices has tampered with very nature of creation. Therefore, there is an urgent need to adopt ecologically friendly, less energy intensive, low cost technologies, instead of depending on high technologies which are sapping out natural resources and farming community.

Brazil experience

Since 1977 the Brazil has taken to carbohydrate farming, by means of plants that produce sugar and starch in large quantities and from which ethyl alcohol is produced as a source of energy, which can be obtained by a simple fermentation. The technology is well developed, which has to be exploited at rural level itself to produce varieties of commodities.

The average yield of sugar-cane per hectare in our country is between 35 to 50 tonnes which represents an annual energy crop equivalent to 25 to 30 tonnes of Coal, 14.5 tonnes of crude oil. In other words 4.5 tonnes of alcohol per hectare can be obtained from sugar-cane plantations. This is done by converting molasses into ethyl alcohol by fermentation.

Further, we have number of native Euperbiacea species which can be cultivated as energy farming species among them *Jatropa curcus* and many latex yielding species can be raised as energy yielding species. Fortunately energy plants differs from food plants in their demands. Most of the species need very little care and inputs and very little water. The only technology that requires to be developed is to evolve genetically superior species for higher production hydrocarbons. Most of the species can be planted at close espacement, with little managements and incorporating the required genetic fabric, the energy production per unit area by biofuels can be increased to a very high level.

Automobile fuel produced from Alcohol, known as *Gashol*, can be used in rural areas, instead of depending on external agencies for supply of energy in the form of Hydropower, Diesel, Kerosene, Petrol etc. Rural industries on fermentation and Enzyme technology or Biocatalysis has to be utilised to produce *Biofuels*.

Karanja (*Derris indica*) is one of important species grown in a agroforestry system. The leaf is an excellent green manure. By using Karanga oil and wood, required energy can be generated by pyrolysis method.

Agroforestry as a new frontier area for economic growth and ecological balance

Strong agricultural economy is a major factor is ushering in a more prosperous and confident future. Success or failure of different land use practices depends largely on the prevailing site quality. Nearly 25-30 per cent of the rainfed agricultural land can be utilised exclusively for forestry, agriculture in such areas will never be a profitable venture. Multipurpose tree species will fulfil the purpose of meeting the varieties of basic requirements like food, fodder, fuel, fertilizer and fibre.

In dry land agriculture, nearly 15-20 per cent of the productive area have been utilised for forming bunds, roads, boundaries, periphery and hedges etc., These areas has to be maintained regularly by incurring a considerable expenditure for repairing particularly, soil conservation bunds, fence roads and boundaries. These areas can be put into productive use by planting in rows or in clusters along the bund that will consolidate the bund and also produce the required biomass as fodder, fuel and other non-wood resources. While planting, care has to be taken to plant only those species that can (1) enrich the soil, (2) the architecture of trees should be such that over story canopy should allow light to penetrate easily, (3) species should not have any allelophathic effect, (4) they should be deep root feeders (5) preferably they should have ability to fix atmospheric nitrogen through symbiotic rhizobium.

ICRAF (International Council for Research in Agroforestry) Nairobi, (Kenya) has prepared a tree data base in which they have carefully selected the species for each planting design, like hedge row, inter-cropping systems etc., assessing the trade offs between the positive and negative effects of trees in each specific environment. They also have a data of 11 thousand species. This list includes species to combine good growth with minimum competitiveness and adoptability to problem sites. Information can be obtained from ICRAF for each region and a suitable native species of similar silvicultural and ecological characters can be recommended for planting in rainfed agricultural areas.

In our country a large number of leguminous species of trees, shrubs, herbs, climbers, creepers and twiners exist. Most of them are multipurpose species.

The species has to be combined to create a vertical competition by close planting and climbers and twiners also to be planted so that they too can trap solar energy and produce required biomass, as fodder food, fibre, and other non-wood resources.

Important tree species are *Acacia* spp. *Albizia* spp., *Sesbania* spp., *Erythrina* spp., *Dalbergia* and many species from Papilionacea sub-family can be grown as companion species.

What is needed is an integrated approach to meet the needs of the region and to develop strong links to industries to make dry land agriculture economical and profitable. Simultaneously technology for process of produces at primary and secondary levels in the rural centres is an imperative to increase the rural employment and to get reasonable price for the produces and to reduce the transport and incidental costs which are much higher in urban centres.

REFERENCES

Anon (1991). Rural energy in the Asia-Pacific region, *Rapa Bulletin*, No. 1, Regional office for Asia and the Pacific, F.A.O., Bangkok, pp. 1-60.

Fonzen, P.F. and Oberholzer, E. (1984). Use of multipurpose trees in hill farming system in Western Nepal, *Agroforestry systems*, pp. 187-197.

Jambulingam, R. and Fernandes, E.C.M. (1986). Multipurpose trees and shrubs on farmland in Tamil Nadu *Agroforestry systems*, 4, pp. 17-32.

Nair, P.K.R. (1987). Agroforestry and firewood production In : Hall, D.O. and Overend, R.P., eds. *Biomass*, Chichester (U.K.), Wiley, pp. 367-86.

Vimal, O.P. (1987). Energy crisis and role of agroforestry resources, In : Agroforestry for rural needs, Khosla, P.K. and Khurana, D.K. eds., I.S.T.S., pp. 289-296.

Chapter 6

Alley Cropping an Alternative to Jhum Cultivation

R.P. Tiwari and L.K. Jha***

In recent years, jhum cultivators are facing acute poverty due to insufficient production of crops with resultant increase in forest areas under jhum cultivation, with shorter rotation period. The reduced period has decreased production potential and created environmental hazards like soil erosion, water loss, loss of fertility, degradation of land and finally threatens the ecological balance of the region.

The decrease in practice of jhum cultivation is happy situation but one question strikes in our mind :

"Are Farmer's Gradually Leaving Jhum Cultivation Wilfully With Great Zeal ?".

In our opinion answer is no. This question should be discussed at length because these farmers are backbone of the state economy. We feel the reason is socio-economic, and biophysical causes of the land. Today, we can see long range of barren hills. The family once dependent on jhum cultivation are now facing problem in maintenance of their family due to various ecological and socio-psychological problem. Now they are migrating in the urban areas in search of jobs.

The present paper, deals, how jhum farmers can be settled by introducing an integrated land use management programme i.e. agroforestry. Successful implementation of this system will help jhumia's in the following ways :-

**Deptt. of Geology, NEHU, Mizoram Campus, Aizawl-796007.*
***Deptt. of Forestry, NEHU, Mizoram Campus, Aizawl-796007.*

* Farmers will get regular sustained income from their farm land and also regular employment.
* Regular employment in their farm land.
* Protection of jhum land from soil erosion, more moisture conservation and fertility status will remain balanced.
* No further extension of jhum land.

What is agroforestry ?

Growing of crop along with tree species is very old practice. To make the subject clear. We would like to discuss Social Forestry first. According to Jha (1991) social forestry is plantation outside the traditional forest area (marginal, sub marginal, private land and community land) and in degraded forest area provided above categories of land are not available, by the involvement of individual below the poverty line or socially or economically depressed, with a view to meeting their requirements in respect of crops, legumes, tubers, medicinal plants, fuel, fodder, timber, etc., Above definition indicates Social Forestry also includes agroforestry. Agroforestry is an integral part of Social Forestry and hence under Social Forestry Programme the production of food crops, legumes etc., are feasible. Lundgrean and Raintree (1983) have stated that agroforestry is a collective name of land-use systems and technologies where woody perennials (trees, shrubs, palms, bamboo etc.) are deliberately used on the same land management unit as agriculture crops and/or animals, either on the same form of spatial arrangement or temporal sequence. In agroforestry system there are both ecological and economical interactions between the different components.

This definition outlines the broad boundaries of agroforestry and the typical characteristics of such a system.

* Agroforestry normally involves two or more species of plants (or plants and animals), at least one of which is a woody perennial.
* An agroforestry system has two or more output.
* The cycle of an agroforestry system is always more than one year.
* Even the most simple agroforestry system is more complex, ecologically (structurally and functionally) and economically, than a mono cropping system. Interested readers can go through

the paper written by Chauhan (1990), Getahum and Rashid (1988), Singh and Pandey (1989), Jha and Singh (1980), Mishra and Ramkrishna (1983), Singh and Singh (1980), Jha and Singh (1991), Kokewe (1990), Cole (1990), Young (1990), etc., for more information.

Characteristics of ideal Multipurpose tree species

While selecting MPTs for agroforestry system, only such species should be introduced which show proper growth, with ability to increase productivity of site. A good MPTs should have following characteristics (Jha, 1992).

1. Production of firewood, timber, green manure, fodder on sustained basis.
2. Fast growing and ability to produce a flush of fresh shoots when cut at ground level.
3. Minimum competition should between crops and multipurpose tree species. The problem like sunlight penetration etc., can be tackled by judicious canopy management.
4. Tree species having nutritious and palatable leaves are preferred, in order to supplement the leaf fodder during lean period.
5. MPTs having ability to fix atmospheric nitrogen in addition to other productive benefits should be preferred.

In selecting the crops to be grown in the interspace attention must be given to the following points.

1. The crop should not deteriorate the soil.
2. Spacing of the main tree crop and interspace should be planned according to their requirements.
3. Intercrop may be utilised as green manure for improving organic matter content of the soil. This will also check weed, control soil erosion and help in enriching nitrogen content of the soil.
4. Combination should be such as to maintain soil nutrients at optimum level and soil erosion can be minimised.
5. Intercrop may be utilised as mulch. This is very common in tropical and subtropical conditions. The mulch acts as a surface barrier preventing evaporation, suppressing weed growth and decreasing soil temperature.

The best mulch will absorb little moisture and permit the rainfall to move downwards rapidly into the soil. This help in improving the soil moisture content and moisture retention capacities.

6. It should also be borne in mind that grain forming timing of the crop should not coincide with flowering, bud differentiation in fruit trees. During non bearing period farmers can grow any crop.

Comparative analysis of shifting cultivation and improved technology

As in now well known, shortening of the jhum cycle does not allow the soil fertility to build up. Thus, farmers need larger areas to be cultivated for the same quantity of produce. Many research centre have been carrying out experiments to tackle the problems of shifting cultivation.

Soil loss, surface runoff and water conservation

ICAR publication (1983) specifically points out that the average soil loss due to shifting cultivation is 40.9 tonnes/ha in a study at Meghalaya. The highest sediment yield however, was 76.9 tonnes/ha/yr. The annual loss of soil was computed by taking into account an average soil loss of 40.0 tonnes/ha/yr. This report also indicates that annual soil loss due to shifting cultivation in the region was about 15.5 million tonnes. Jha and Rathore (1981) mention that erosion ratios of Orissa soils under shifting cultivation were 18.03 and 18.78 in the surface and 6.5 and 11.4 in the subsurface as compared with erosion ratio in soils not under shifting cultivation of 10.7 and 10.8 in the surface and 3.7 and 11.7 in subsurface soils. Erosion ratios were higher in the upper layer than in the lower layers. Soils under shifting cultivation had high erosion ratios. While soils not under shifting cultivation were within the safety limits of erodibility. Borthakur *et. al.*, (1978) studied that surface run-off under shifting cultivation was about 11.40 mm. Partial terracing reduced surface run-off further to 32.8 mm. This study indicates that physical barriers considerably reduces surface run-off.

In recent decades, continuous cropping have accelerated the soil erosion and condition is progressively deteriorating. In most of the shifting cultivation areas wrong soil tillage practices also play important role in enhancing the rate of run-off and soil erosion. In these areas luxurious vegetation mostly (Grass species) makes people to consider, that the soil is extremely rich, but this impression is misleading. In fact soils are poor and heavy rain destroy them within few months or few years.

Young (1989) mentions that trees are capable to control erosion. This is achieved in two ways with trees acting as barriers or as cover. The barrier function in the conventional approach to erosion control by checking run-off of water and the canopy reduces the raindrop impact. On slopping land, leguminous tree species when planted along the contour, minimises soil erosion. Tree foliage can be used as mulch and fertilizer for food crops (Reynolds *et. al.*, 1988).

Nutrient status

Mishra and Ramkrishna (1983) have studied the effect of jhum on soil fertility at high elevations of Meghalaya using 15, 10 and 5 years jhum cycle and terrace system. Soil nitrogen concentration under a 5 years cycle was significantly lower than under the 10 and 15 years cycles. Nutrient declined sharply after the burn in the surface layers and was attributed to volatilization. The degree of volatilization is dependent on the intensity of the burn and, therefore, the nitrogen decline is lower in a 5 years cycle than in longer cycles. Nitrogen concentration in the soil improved at the end of one year under all cycles, but such an improvement was observed only under 10 and 15 years cycles.

Under terrace cultivation, nitrogen losses continued up to the end of two years. Available P was generally lower under a 5-yr cycle than 10 and 15 year cycle. Available P-under the three year jhum cycle declined significantly after the burn. Recovery occurred after 30 days under 10 and 15 years cycle and after 100 days for the 5 years cycle with subsequent decline during cropping. In the terrace system, the available P declined during cropping, after an initial increase. The concentrations and total quantities of all cations on the surface soil layers under the three year. Jhum cycles improved markedly after the burn owing to releases through ash, but subsequently declined sharply. The declined is partly due to absorption by the developing crop. Large quantities of cations were released under 10 and 15 year jhum cycles owing to greater quantities of slash burned than that under a 5-yr cycle. Starting from a lower level in a 0-year fallow, the recovery of C, N and available P occurred in older fallow of 5 and 10 years with a slight decline in a 15 years fallows, longer fallows greatly improve humus and nutrients to sustain slash and burn agriculture. A jhum cycle of 5-years is inordinately short for nutrient recovery. Longer cycle of 10-years is the minimum requirements. A terrace system does not offer a better alternative to short cycle because it cannot sustain continuous cropping without heavy fertilizer inputs. Chauhan (1990) citing the results of report of ICAR Research Complex

(1983) has concluded that in slash and burn practices organic matter of the surface soil is also oxidised. This may cause a decrease in soil organic matter. The decrease in organic matter is more pronounced when the temperature during burning exceeds 150°C. Both erosion and cultivation caused a decrease in the organic matter content of the soil. The annual loss of soil and crop yield due to the shifting cultivation in North Eastern region was estimated to be 15.5 million tonnes and 52.23 thousand tonnes respectively (Chauhan 1990). The on-site cost of soil degradation was computed to be Rs. 980.6 ha/yr which was 36.6 per cent of the value of the harvested crop produce. On regional basis, the on-site cost of soil degradation amounted a loss of Rs. 37.9 crores/yr. The loss crop yield, plant nutrient and land degradation contributed 48.2, 39.6 and 12.2 per cent, respectively towards the on-site cost of soil degradation. Among the plant nutrient nitrogen loss is the maximum (29.2) followed by phosphorus (7.8%), potassium (1.6%) and sulphate (1.0%).

Under agroforestry systems, hedgerows may be utilised as green manure or mulch, which improves soil nutrient and moisture. Under alley cropping system moderate fertilizer or F.Y.M. may be best solution to yield stability and increased production under modified shifting cultivation models.

In the jhum cultivation, villagers cut and clear a strip of forest three to four meters wide all around the jhum plot as fire line before they set fire to the area and scrupulously guard against fire spreading outside the area. The control on protection of forest is gone and now more damage to vegetation is caused by escaping jhum fire than the actual practice of jhum cultivation itself. The sprouting young vegetation covering the left over jhum area is thus destroyed year after year till the land becomes completely devoid of vegetative cover worth the name (Status report, AFCL, 1989). In addition, increase in temperature also destroys beneficial micro-organisms. But, there are some positive attributes. The smoke liberated in the slash and burn cultivation, kills pathogens and pests.

A look into future

Before implementation of modern technology, we should study in depth the psyco-socio-cultural profiles, particularly the social organisation, social behaviour and social cognitions associated with shifting cultivation and permanent settlement patterns by employing standard psychometric and sociometric parameters. The finding of above study would provide empirical foundation to formulate strategies for cognitive and behaviour modification of individuals and community orientation programmes for planned elimination minimization of jhuming and to facilitate permanent settlement patterns relating to agriculture, forestry and animal husbandry. Therefore, meeting this objective means rehabilitation of the shifting cultivation area for the rehabilitation of the people.

It is apparent from above that the approach of jhum cultivation should be need based and location-oriented. This practice can be only modified by studying existing agriculture practices and utilising its natural resources, keeping in mind the psyco-socio-economic problems of the area.

Several design models have been suggested to solve the problems of shifting cultivation. The three tier system suggested by the ICAR, Shillong i.e. cultivation of trees at the hill top, cultivation of horticulture crops in the middle portion of the hill slopes and forming terraces at the lower portion of the hill slopes and at the lower portion for cultivation of field crops. Pandey and Singh (1984) have suggested the tree rotation shifting cultivation. Each of these methods as an alternative to shifting cultivation has limitations.

In our opinion any alternative models can only be adopted by local people if it suits local environmental condition and meet the needs of farmers. Besides this, the model should be simulate traditional practices with minor alternations. The above problem can be mitigated by developing appropriate agroforestry models for each agroclimatic zone (different altitude. aspect etc.). Introduction of all the three components i.e. agronomy, forestry and animal husbandry will help in minimising the soil erosion, conservation of moisture, increase in infiltration rate, decrease in evaporation from soil, balance nutrient status and finally help in integrated use and sustained land productive. It may be stated that in agroforestry one has to depend on on-station experiments, where conclusive results can be obtained after 10 years. We generally try to transfer the technology in the farmers field without considering biophysical and psyco-socio-economic problems of the farmers. Farmers never accept imposed technology under above circumstances impact remains for short period. In my opinion we should not spend time in on-station trial. Our extension worker/scientist should work along with farmers to examine requirements of the farmers and bio-physical parameters. On-farm trials (in the farmers field) in collaboration with farmers should be tried on model basis in 5 or 10 village in 1st phase. Here again I would like to mention that this practice cannot be successfully introduced by any single department. To conduct this programme personnel of agricultural, forestry, soil conservation, social science and animal husbandry department should work under one umbrella.

Thus, proposed alley cropping models on agri-silviculture, agri-silvi-horticulture or agri-horticulture practice may be best solution for rehabilitation of jhum land.

Suggested (Alley Cropping) models for the rehabilitation of jhum land

I have not discussed about the slope and nature of crops to be tried in particular zone, because farmers generally practice jhum cultivation within permissible level of slope. Degree of slope will help in computing the alley width under agroforestry system. Extension workers and scientists should be careful in computing the alley width. Agricultural crops, tree species vary in different agro-climatic zone. In each agro-climatic zones models must be tried with local species as well exotics. The proposed model of alley cropping systems are as follows :-

Model-1 : Intensive hedgerow (alley) cropping on the jhum land

Objective

* This model improves fertility status and other properties of soil.
* Conserve more moisture and cause less soil erosion.
* Maximum utility of site by increasing biomass productivity.

Alley cropping involves planting of rows of trees or shrubs, along contours, and food crops and cover crops are grown in between the rows. The trees will limit soil erosion, decrease rate of run-off and tree foliage can be used as mulch or green manure for agriculture crops or for animal feeds. Tree foliage is used to maintain and improve soil fertility, while cropping is in progress, similar to natural processes, that occurs during the traditional fallow period, thus increasing land productivity.

Perennial vegetation (hedge forming species)

(1) *Leucaena leucocephala* (K8, K28, etc.,)

(2)* *Gliricidia* species

(3)* Perennial pea

(4)* *Vetiveria* species

(5)* Any other local Nitrogen fixing tree species suitable for agroforestry system.

Pruning management

* At different lopping intensities.

Alley width = 6 M or may be changed according to degree of slope of the jhum land.

Plot size = 15 x 6 M (may be changed).

Treatments

T1 = Perennial vegetation (1-5No) + Agricultural crop two or more/vegetables two or more.

T2 = Perennial vegetation (1-5No) + Agricultural crop two or more/vegetables two or more (Tree foliage as mulch).

T3 = Perennial vegetation (1-5No) + Agricultural crop two or more/vegetables two or more (Tree foliage as green manure).

T4 = Perennial vegetation (1-5No) + Agricultural crop two or more/vegetables two or more (Farm yard manure as fertilizer).

T5 = Perennial vegetation (1-5No) + Agricultural crop two or more/vegetables two or more (Nitrogen application).

T6 = Agricultural crop two or more/vegetables two or more (Nitrogen application).

T7 = Agricultural crop two or more/vegetables two or more (F.Y.M.)

T8 = Agricultural crops/vegetables - control (without any treatment)

T9 = Tree species alone - control.

Parameters

* Yield of annual crop.

* Nutrient status of soil.

* Nitrogen fixing capacity of the tree involved.
* Growth performance of tree species.
* Percentage decrease in soil erosion, run-off and increase in moisture retention capacity of soil.

Model-II : Alley cropping with perennial grass

Objective

Utilisation of nitrogen fixing tree species with grasses to improve the soil fertility, reduce soil erosion and improve the production of fodder in previous jhum land.

Hedge forming species (H.F.S.) :

* *Leucaena leucocephala*
* *Gliricidia* species
* Perennial pea
* *Vetiver* grass
* Any other local Nitrogen fixing tree species.

Grass species

Two or more grass species suitable for the particular agro-climatic zone.

Pruning management

At different lopping intensities and intervals

Alley width

5M or may be changed according to degree of slope.

Plot size

15 x 5 (may be changed).

Treatments

T1 - H.F.S. (1 to 5 No.) + Grass or more species.

T2 - H.F.S. (1 to 5 No.) + Grass two or more species (Tree foliage as mulch).

T3 - H.F.S. (1 to 5 No.) + Grass two or more species (Tree foliage as green manure).

T4 - H.F.S. (1 to 5 No.) + Grass species (nitrogen application).

T5 - H.F.S. (1 to 5 No.) + Grass species (F.Y.M.)

T6 - Grass alone.

T7 - Tree species alone.

Parameters

* Yield of grass
* Growth performance of tree species
* Nutrient status of soil.
* Soil erosion, run-off and moisture contents of the soil.

Model-III : Introduction of agril. crops/vegetables species under different spacing of tree species including fruit trees on the jhuming land or degraded jhum land.

Objective

* To find out best combination of tree species and intercrops.
* To study impact of trees/intercrops on the fertility status of soil.
* To reduce soil erosion and increase water conservation.

Tree species to be tried

* Important specially accepted tree species.
* Important fruit trees.

Inter crop

* Agril. crop - Two or more
* Vegetables - Two or more (Rabi & Kharif season)
* Spices - Two or more

* Fruit - Pineapple, sweet potato etc.,
(Selection of species depends upon farmers choice and agro-climatic zone).

Spacing

3 x 3 M, 4 x 4 M, 5 x 5 M or may be changed according to degree of slope.

Plot size

24 x 24 M, 12 x 12 M (May be changed).

Treatment

T1 = Tree species + Agril. crop two or more/vegetables two or more/ spices two or more/fruit species two or more (control).

T2 = Tree species + Agril. crop two or more/vegetables two or more/ fruit two or more species (nitrogen application).

T3 = Tree species + Agril. crop two or more/vegetables two or more/ fruits two or more species (F.Y.M. application).

T4 = Tree species + Agril. crop two or more species/fruits two or more species/vegetables two or more species (Mulching by natural grass).

Parameters

* Growth performance of tree species.
* Yield of intercrops.
* Nutrient status of soil.
* Increase or decrease in soil erosion, run-off and moisture content of soil.

Model-IV : Horti-Silvi-Agricultural system (wide alley cropping) in the jhuming land.

Objective

* To meet the requirement of fruit, wood crop and the fodder.
* To conserve more moisture and reduce soil erosion.

* Maximum utility of site by increasing productivity. (Fruit trees, nitrogen fixing hedge forming species or vetiver grass may be integrated to meet the diverse requirement of the farmers. The space between fruit trees will be utilised by *Leucaena leucocephala* or any other N.F.Ts. as fillers to form hedge).

Spacing

5 x 5 M (may be changed according to percentage of slope).

Plot size

30 x 30 M (may be changed).

Treatment

T1 = Fruit tres (Citrus/Aonla/Guva etc.) + Hedge forming trees species as fillers between trees + Agril. crops two or more/ vegetables two or more fruits two or more (Ananas/Banana etc.) (Control).

T2 = Fruit trees + Hedge forming trees species between trees + Agril. crop two or more/vegetables two or more fruits two or more. (Hedge foliage as mulch).

T3 = Fruit trees + Hedge forming trees species + Agril. crop two or more/vegetables two or more/ fruits two or more. (Hedge foliage as green manure).

T4 = Fruit trees + Hedge forming trees species + Agril. crop two or more/vegetables two or more/fruits two or more. (Nitrogen/ F.Y.M. application and hedge foliage used as fodder).

T5 = Agril. crop/vegetables/fruits (control).

T6 = Fruit trees (control).

Parameters

* Growth performance of fruit tree species.
* Yield of intercrops.
* Nutrient status of soil.
* Soil erosion, run-off and moisture contents of soil.

Conservation of soil and moisture and maintenance of fertility status under the proposed alley cropping models.

At present forests are disappearing at a faster rate due to jhum cultivation. Jhum farmers prefer mono cropping. Generally they do not care growth performance or regeneration ability of tree species within jhum land. Jhumias life style is thus, going to disturb normal environmental equilibrium and jhumias do not get required tree products easily.

Jhum land are left barren with concomitant increased rate of run-off and soil erosion. The accelerated soil erosion influences water regime within its range.

In some parts, progressive farmers form soil bund in their plots and a channel for the run-off. Bund is made up of soil of the same plot. After heavy rains, bund soil gets eroded within a year. This measure brings short term relief.

Some farmers use gunny bag filled with soil to form terrace or stones or branches of trees are used to form natural wall across the bed. The height is raised above the level of field. This structure reduces soil erosion, silt gets deposited near the stone wall and water oozes out. This structure also provides increased infiltration of the rain water that fall in the crop field. This is a good and an effective method to control soil erosion for short period and helps in the formation of natural terrace. After few years, this structure also looses its usefulness, fails to check soil erosion and run-off etc.

We have discussed all the traditional approach as that helps in solving the problems of shifting cultivation for a short duration.

The problems of jhum land i.e. soil erosion, increase run-off, loss in nutrient status can be solved by adopting suitable agroforestry alley cropping models. My suggestion is to introduce intensive hedgerows or alley cropping or wide alley cropping system. Horizontal distance between the hedge depends upon the slope percentage. The distance between the hedge decreases with the increase in slope percentage. The most important features of this system is that farmers can do the whole work himself. Labour requirements is less. The hedgerows must be establish across the slope on the contour line. Contour line passes round the slope and it is line of equal elevation. Hedgerows when established in straight line, will increase the hazards of shifting cultivation. Under intensive hedgerows system nitrogen fixing tree species e.g.. *Gliricidia* species, *Leucaena leucocephala* or any N.F.Ts. suitable for agroforestry system are planted at close spacing. To make barrier effective, trees should be spaced densely 5-6 cm with in rows. Two rows may be considered as t0he best. If nitrogen fixing tree species are planted at larger gap, gap between trees facilitates water flow with increased soil erosion, nutrient loss. To overcome the above difficulties the space between the trees are used for perennial grass or pigeon pea or vetiver grass etc. If farmers are willing to plant fruit trees or tree species having longer rotation period (timber species). Naturally spacing between the trees will be more. In such condition *Leucaena leucocephala*, vetiver grass, any suitable perennial grass or pigeon pea, may be used as fillers within the rows. The hedge forming tree species, grasses should be harvested after proper establishment. Within two years, hedge forming species will have well established root system, after that hedge species may be managed under coppice system. Hedge forming, tree species are generally felled above 10-15 cm from the ground. Vetiver or perennial grass can be harvested down to ground level.

In the initial year, behind each hedgerows, a contour trench may be formed parallel to the hedgerows, so as to serve as soil traps. After establishment of the hedgerows soil is filtered out of the run-off and run-off oozes through the hedge formed on the contour. Silt are deposited behind the plant and water flows on the slope with slow speed with very low quantity of silt mass. In this way natural terrace are formed. In this system farmers has to take special care of hedge in the initial year till the establishment of natural hedgerows at the hedge of terrace with deep root system.

The small twigs and foliage of (nitrogen fixing tree species) hedgerows may be incorporated in the soil and branches may be used as fuelwood. The micro-organism in the root nodules of nitrogen fixing tree species fixes atmospheric nitrogen, which assists in better growth of the inter crops. Leaves when used as green manure, release humus and minerals to the soil after decomposition. The roots of nitrogen fixing tree species due to nutrient pumping action carries important nutrient from the lower zones of soil and makes it available to agricultural crop. In the regular coppicing management, roots die and decompose and add organic matter to the soil. In the jhumland, hedgerows hold soil together along the contour. The hedgerows also help as windbrakes. The foliage of NFTs can be utilised as fodder or mulch. This system will improve soil nutrient status similar to land kept fallow for two to three years in the jhuming areas. More moisture will be stored in the field, which in turn will help in better growth of intercrops. To protect the intercrops from shade effect, pruning may be done as required.

We can conclude that this system is highly flexible. Farmers can modify the model as per their needs, environmental, social and economic condition, and it can protect jhum lands from degradation and may remain productive on a sustained basis. These models are only technical upgradation of the traditional practices.

Suggestions

(1) Government is attempting to motivate the jhum cultivators to settled scientific cultivation. In the hilly areas there is no shortage of land till date. Government should allot land permanently to the jhumia's, individually or on community basis, so that they can think that permanent reclamation of jhum land is in the interest of their own development and they may not get enough new areas for this practice.

(2) Establishment of a network of alternative models in each zone in conformity to traditional practices which should be demonstrated to the jhumias.

(3) As the system is multidisciplinary, separate education and training facilities for development, management and extension of agroforestry systems at different level are required.

(4) Most of jhumia's are below the poverty line. Moreover, in spite of the efforts made during five years plans, infrastructure to replace in our jhum belts is not up to the mark. The facilities for education, drinking water, transport and communication, institutional credit etc., are lagging far behind. Under these circumstances we cannot expect voluntary involvement of jhumia's to settle scientific cultivation.

(5) Vigilance cell consisting of jhumia's representative, should be formed at remote villages, so that jhumia's may avail themselves of timely benefits granted by the Government. They must be protected from the practice of greazng palm of corrupt officials.

(6) Frequent transfer of senior officers, who are directly involved in the development programmes including rehabilitation of jhumias should be discouraged as it adversely effects the execution of rehabilitation programmes.

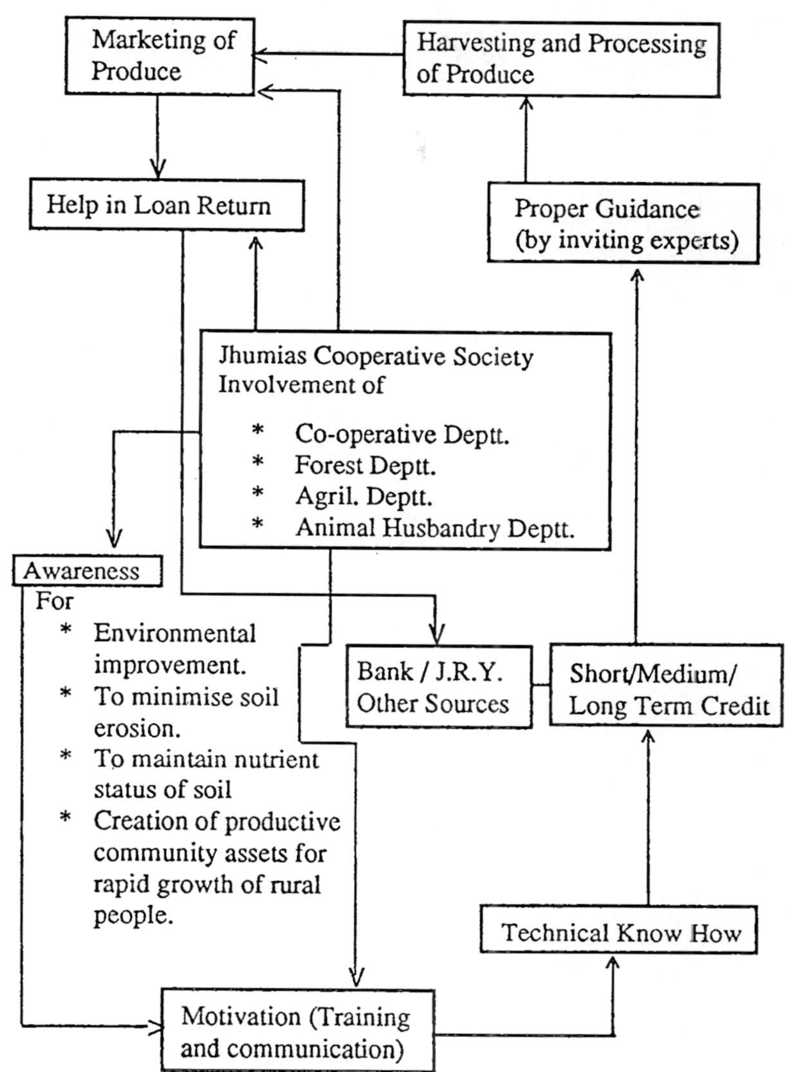
Marketing of Produce
Harvesting and Processing of Produce
Help in Loan Return
Proper Guidance (by inviting experts)
Jhumias Cooperative Society Involvement of
* Co-operative Deptt.
* Forest Deptt.
* Agril. Deptt.
* Animal Husbandry Deptt.
Awareness
For
* Environmental improvement.
* To minimise soil erosion.
* To maintain nutrient status of soil
* Creation of productive community assets for rapid growth of rural people.
Bank / J.R.Y. Other Sources
Short/Medium/ Long Term Credit
Technical Know How
Motivation (Training and communication)

Chart 1

(7) *Jhumias co-operative society* : Needless to say co-operatives can play a pivotal role in accelerating the tempo of rehabilitation of jhum land or jhumias. The jhumias are the weaker sections of our society. It is assumed that proper functioning of jhumias co-operative society would reduce their economic exploitation and raise their socio-economic status. Presently utmost emphasis is laid for the rehabilitation of the jhum land to increase productive and sustainability of jhum land. Co-operatives will also improve overall quality of life of jhumias.

The co-operative society will create awareness among the farmers and motivate them to implement rehabilitation programme of jhum land. The society will co-ordinate technical experts/extension workers to guide and provide latest technical know how for successful rehabilitation of jhum land. The jhumias co-operative society will arrange timely short, medium and long term of credit. Credit will be provided for consumption purpose during the lean period of the year to eliminate exploitation by money lenders or in the initial years of settled cultivation. The jhumias will get technical guidance from the expert through society. Marketing of produce will be exclusively done by primaries. Consequently, exploitation by the intermediaries will be eliminated the processing and marketing of the produce by the society will also facilitate the timely recovery of loans and hence the problem of loans remaining outstanding will be automatically solved. Thus it is apparent that co-operative society can bring multiple benefits to jhumias, solve several intractable problems and help in improving the overall quality of life of jhum cultivators and help in the formation of egalitarian social order.

(8) Thus in our opinion to reclaim jhum land, 1st phase is permanent allotment of land to jhum cultivators and 2nd phase is creation of awareness, motivation imparting technical know-how, financial assistance etc., to jhum farmers. Farmers may be provided these incentives through co-operative societies or any other suitable agencies. The farmers can utilise available space and resources by adopting an appropriate agroforestry model. Suppose farmers have been allotted 10 hec. Land planning should be such that after construction of house, kitchen, pig and cowshed, compost pit, Gobar gas plant, only tree species to fulfil

the requirements of fuelwood or silk host plant may be planted. This will also assist in income generation of farmers after 4 years and lops and tops of plants can be used as domestic fuelwood. Farmers can put their third plot under multicropping system. Fruit trees at larger spacing 12M x 12M or as suggested by experts, in between the papaya, banana or any other fruit, spices may be planted. In the interspace agricultural crop, spices vegetables etc., can be grown as per their own domestic requirements and demand of local market. Farmers can grow perennial flower or seasonal flower in a small part of his field and can practice apiculture to increase their income. This will also augment natural enemy complex and improve biological control of pests. The hedgerows foliage can be utilised as green manure or fodder. As discussed earlier under Govt. financial help, farmer should be provided cow/pig/goat etc. The grass species and foliage of nitrogen fixing tree species would provide fodder. The farmers can use farm yard manure of his compost pit to improve fertility of his crop field and rearing of cow/pig/goat etc., will supplement income. It is apparent that under agroforestry systems involvement and co-ordination of all the three department i.e. Agril., Forestry and Animal husbandry are essential. Finally, this system will improve the economic status of jhum cultivator and loan provided by the Govt. financial institution can be recovered in instalment. In this initial year jhum cultivators will maintain their family from the loan provided by the Govt. Concerned efforts are to be made by all to enthuse jhumias for voluntary mass involvement, so that, economic stability may be ensured. (Jha, 1992).

REFERENCES

A.F.C.L. (1989). Status report on Dhaleswari catchment Mizoram, Department of Agriculture, Government of Mizoram, pp. 1-200.

Borthakur, P., *et al.* (1978). Cited by R.K. Gupta, (1981). Ecological consequences of irrigation land uses in loss of productivity soil, In: South east Asia Regional symposium on problems of soil erosion and seditation, (T. Tingsanhall and H. Eggers, eds.) Bangkok, Thailand, pp. 27-29. (C.R.)

Chauhan, B.S. (1990). On site degradation due to shifting cultivation in North Eastern Region NEHU, *Journal of Social Sciences and humanities,* Vol. VIII, No. 3, pp. 5-8.

Cole, Z.A. (1990). Cassia alley cropping in the Gambia. *AFNETAN News letter*, Vol. 2, No. 2, pp. 6-7.

Getahum, A. and Rashid, K. (1988). Agroforestry in Kenya, A field guide, Rural afforestation extension, Forest Department, Ministry of Environment and Natural Resources in collaboration with the Swiss Development Co-operation, pp. 1-59.

I.C.A.R. (1983). Shifting cultivation in North Eastern India, Research Complex for North Eastern Region Shillong, Meghalaya, India, pp. 35-40.

Singh, J. and Pandey, M.C. (1989). Some measures to improving shifting cultivation, NEHU, *Journal of Social Sciences*, Vol. V, No. 4, pp. 9-12.

Jha, L.K. (1991). Social/agroforestry as an important component in the development of tribal villages. A case study. Mendal, 6(3-4) pp. 487-490.

Jha, L.K. (1992). Agroforestry - A stable alternative to jhum cultivation, Int. seminar, *studies on the minority nationalities of North Eastern India*, Directorate of higher technical education, Govt. of Mizoram, Aizawl, pp. 129-136.

Jha, M. and Singh, M. (1980). Effect of various stages of shifting cultivation on soil erosion from steep hill slopes. *Indian Forester*, Vol. 107(5), pp. 310-313.

Jha, M. and Rathore, R. (1981). Erodibility of soil in shifting cultivation areas of Tripura and Orissa, *Indian Forester*, Vol. 107(5), pp. 310-313.

Kokewe, M. (1990). On farm agroforestry trial in Luapula province Zambia, *AFNETAN News letter*, vol. 2, No. 2, pp. 6-7.

Lundgrean, B. and Raintree, J.B. (1983). Sustained agroforestry in Agricultural Research for Development : Potential and Challenges in Asia, (B. Nestel, Ed.), The Hague, ISNAR, pp. 37-39.

Mishra, B.K. and Ramakrishnan, P.S. (1983). Slash and burn agriculture at higher elevation in North Eastern India. 11. Soil fertility changes *Agric. Ecosystem and Environment*, pp. 83-96: Cited by R.E. De cruz and N.T. Vergara, Protective and ameliorative role of agroforestry, An overview, *Agroforestry in the humid tropics*, pp. 18-19.

Reynolds, L., Atta-krah, A.N. and Francis, P.A. (1988). Alley farming with livestock guidelines. Humid zone Research site, *International Live Stock, centre for Africa,* Nigeria, pp. 1-30.

Roy Burban, B.K. (1989). Critical appraisal of shifting cultivation farming technology and of various approaches, *Journal of N.E.C.*, Vol. X, No. 2-3, pp. 1-4.

Shanan, D.A., Kabaluapa, K.N. and Ngoyi, M.L. (1990). Alley cropping stabilizes maize yields in Gandhijika, Zaire, *AFNETAN*, Vol. 2, No. 2, pp. 8-9.

Singh, A. and Singh, M. (1980). Effect of various stages of shifting cultivation on soil erosion from steep hill slopes. *Indian Forester,* Vol. 16(2), pp. 116-121.

The State Forest Report (1989). Government of India, F.S.I., 25 Subash Road, Dehra Dun (U.P.).

Young, A. (1989). 10-hypothesis for soil agroforestry research, *Agroforestry today,* Vol. 1(1), pp. 13-16.

Young, A. (1990). Agroforestry environment and sustainability, *Outlook on agriculture,* Vol. 19, pp. 155-160.

Chapter 7

Silvipastoral Practices and Conservation of Fodder

*S.P. Arya**

India possesses an enormous livestock population (700 million) but the production of livestock products is too meagre to meet the demand of fast increasing human population of more than 820 million. For the success of livestock production, breeding, feeding and health cover of animals require adequate attention. To improve the production potential of the existing livestock, appropriate breeding programmes have been launched since long with resultant increase better animals. It is well known that these improved breeds of livestock, unless fed properly, will not be able to portray their acquired genetic potential even if reared under an excellent health environment. This low productivity of livestock is a matter of concern, which is mainly due to poor fodder and feed resources. The fodder production at present is too insufficient to meet the requirements of the livestock population and also the forages produced are of poor quality. A recent estimate indicates that the deficiency in the total forage need is about 52 per cent of dry roughages and about 68 per cent of green fodder. In India, only 4 per cent of cultivated land grows forage/fodder because farmers prefer to grow cereal and cash crops. The opportunities for increasing the area under cultivated forages are remote, because of preferential need for human food. Besides, the country is also passing through an energy crisis with the acute shortage of fuelwood which is often substituted by burning cow dung cakes thus, depriving the cultivated land much needed precious organic manure, for increased crop production. Such a situation, therefore, demands full exploitation of the area under cultivated forages, area under farm bunds and roads, permanent pastures, fallow and cultivable wastelands and best use of the available feed and

* *Department of Silviculture and Agroforestry, Dr. Y.S. Parmar University of Horticulture and Forestry P.O. Nauni, Solan (H.P.)*

fodder resources of the country so that adequate good quality nutritious fodder may be available to our livestock and also to meet the fuelwood requirements. It is, therefore, imperative not only to improve the quality of feed and fodder, by integrating new crop cultivators and appropriate conservation technology but also to search for alternate source of animal feeds and further increase and improve resources. The role of fodder trees and shrubs in providing highly nutritious green leaf fodder is of great importance to livestock production, especially during the crucial lean periods when grasses are either grazed or become very rough and unpalatable and therefore, silvipastoral system is superior in terms of forage production, forage quality and period of forage availability and minimizes the seasonal variations in nutrient availability (Deb Roy and Pathak, 1983; Deb Roy, 1988, 1989; Singh, 1989; Saha and Gupta, 1987). While making efforts for increasing the production of fodder and fuelwood, one should bear in mind that the cost of production should be economical. Silvipastoral system of farming has been recognised as a low input technology for increasing forage and fuelwood production from the vast wastelands and degraded grazing lands of the country.

As is well known, silvipastoral system integrates woody species (trees and shrubs) with grass or grass legume mixtures simultaneously or sequentially on the same piece of land. This system involves species that can be grown at different canopy heights and managing them in a manner to obtain maximum benefit through efficient utilization of solar energy and other inputs/resources. The objective is that the system should not only combine system components (trees, grasses and legumes) for their complimentary but should also serve to improve the productivity of the site and to optimise the combined production (Singh, 1989). In recent years the wastelands and the degraded grazing lands have been stripped off the useful and adequate vegetative cover for livestock grazing and are prone to soil erosion. Silvipastoral system of forages and fuelwood production offers a solution by higher production and conservation of these lands. The productivity from improved grasslands through silvipastoral system at Jhansi, yielded a three fold increase (Hazra, 1989).

For the success of this system, selection of production components is important *i.e.* selection of the woody species which form the upper canopy layer must suit to specific ecosystem and the need based production. Similarly the choice of grass and legumes depends on their complimentarity and not competitiveness. *Grewia optiva* is a very important fodder tree of mid Himalayas and provides highly nutritious green fodder during November-December.

Arya (1988a) recorded the leaf fodder, fibre and fuel wood yields *of Grewia optiva* in relation of different girth classes (Table 7.1). In another study (Arya, 1989) it was observed that there is no adverse effect on milk yield, its composition and health of the animals even if 100 per cent digestible crude protein (DCP) is substituted by *Grewia optiva* fodder in the ration of lactating cross-bred cows. Sharma *et. al.*, (1988c) studied the effect of girth and plant height on leaf fodder production in *Bauhinia variegata* and *Robinia pseudoacacia* (Table 7.2). The influence of tree species on the grasses are well marked. The canopy structure and the type of tree species greatly influence the grass production underneath. Arya (1989d) recorded mean dry grass production of 81.95 q/ha under poplar plantation at 4 x 7 m spacing as compared to 55.69 q/ha in 3 x 3 m spacing, whereas under *Eucalyptus* trees planted at 3 x 3 m spacing yield was only 14.92 Q/ha. Sharma *et. al.*, (1988 a) also reported significant effect of tree canopy on the green grass yield (Table 7.3). It is evident from these findings that scientific exploration of tree component in such systems is important for providing sustained fodder and fuelwood yields during lean period and reducing canopy cover to facilitate penetration of light to ground flora.

Table 7.1 : Effect of girth and plant height on leaf fodder and fuel wood production of Grewia optiva.

Girth Class (cm)	Plant height (m)	Leaf fodder yield Kg/ plant	Fibre yield Kg/Plant	Fuelwood yield Kg/ plant
0 - 40	5.977	3.625	0.180	1.6775
40 - 80	5.88	17.375	0.217	7.702
80 - 120	8.907	19.625	1.257	8.657
120 - 160	9.57	46.75	2.282	22.327
160 - 200	10.45	50.75	3.200	23.047
200 and above	9.08	75.137	3.837	35.198
CD 5%	2.509	18.233	1.342	10.004

Table 7.2(a) : Effect of girth and plant height on leaf fodder production in Bauhinia variegata

Girth Class (cm)	Plant height (m)	Green leaf fodder Kg/plant
0 - 20	2.4	0.254
20 - 25	4.04	0.506
25 - 30	5.45	0.824
30 - 35	5.55	1.166
35 - 40	4.48	1.610
40 - 45	7.66	2.236
45 - 50	6.05	4.074
50 and above	6.95	4.098
CD 5%	2.745	1.290

Table 7.2(b) : Effect of girth class and plant height on leaf fodder yield in Robinia pseudocasia

Girth Class (cm)	Height (m)	Leaf fodder yield Kg/plant
0 - 20	6.10	1.250
20 - 25	6.66	2.410
25 - 30	7.30	3.550
30 - 35	4.68	3.936
35 - 40	7.56	4.750
40 - 45	10.53	5.591
45 - 50	8.35	6.172
50 - 55	9.42	2.067
55 - 60	9.47	5.842
60 - 65	9.43	6.658
65 -	7.75	4.177

Table 7.3 : Effect of Khair tree on green grass yield

Treatments	Green yield (Q/ha)
T_1 50 cm away from the tree	72.22
T_2 100 cm away from the tree	72.62
T_3 200 cm away from the tree	80.33
T_4 400 cm away from the tree	82.22
T_5 Open area	83.80
CD 5%	8.67

In a silvipastoral study, Sharma *et. al.*, (1988 b) reported mean grass production of 7.6 t/ha under *Robinia pseudocacia* compared to 5.2 t/ha under *Bauhinia variegata* and 3-5 t/ha in natural grassland (Table 7.4) and concluded that tree component in natural grassland improved the grass productivity in addition to availability of tree leaf fodder and fuelwood. In a forage production trial of *Cenchrus ciliaris* and *C. setigerus* in association with *Acacia tortilis* and *Leucaena leucocephala* at Jhansi it was observed that, *Cenchrus ciliaris* recorded higher forage production in association with *Leucaena leucocephala* (Deb-Roy *et. al.*, 1980). Forage production in this system is also significantly influenced by amount and distribution of rainfall (Deb-Roy *et. al.*, 1989) and different tree densities. This is because of the inherent branching habit of some trees from the base and also of the spreading crown necessitating canopy opening through lopping.

Table 7.4 : Effect of tree species on grass yield

Treatments	Plant height (cm)	Grass yield (Q/ha)
Robinia pseudocasia	77.84	76.6
Grewia optiva	76.93	63.5
Bauhinia variegata	62.85	52.5
Albizia lebbek	88.6	67.3
C D 5%	9.67	N S

There had been very little efforts on systematic studies on pasture production in hills under silvipastoral systems. However, certain available general informations and technology generated at different places in the country are of use for the improvement of the hill grasslands and good deal of work is in progress at Solan. It has been observed that in the lower and mid hills, planting of *Leucaena leucocephala*, *Sesbania sesbens*, *Grewia optiva*, *Celtis australis*, *Morus alba* in the gradonies and introducing improved grasses and legumes viz. *Setaria* and *Dolichos lablab/Siratro* in between or within gradonies have yielding sustainable green fodder from these degraded grasslands. Besides these, some other trees, grass and legumes have been found useful (Table 7.5).

Table 7.5 : Tree species/grasses and legumes for different agroclimates of H.P.

1. *Low hill region (below 800m altitude) and Mid hill regions (800 m to 2000 m altitude.)*

1. Grasses	-	Cenchrus ciliaris, Dicanthium annulatum chrysopogon montanus, Pannicum maximum, setaria (Nandi/Narok/Kanjangula), Hybrid napier, Themeda anathera chrysopogon fulvus Arundinella nepalensis, Bothriochloa pertusa.
2. Legumes	-	Siratro, stylosanthes hamata, Dolichos leb-leb, velvet bean clitoria ternatia, white clover, green/silver leaf desmodium.
3. Trees	-	Leucaena leucocephala, Grewia optiva, Morus alba, Celtis australis, Bauhinia varigata, Albizia lebbek, A. Amara, A. procera.

2. High altitutde region (2000 m and above)

1. Grasses	-	Dactylis glomerata, Phleum pratense, Festuca pratense, Phlaris tuberosa.
2. Legumes	-	White clover, Red clover, Lucerne, Lolium perenne.
3. Trees	-	Robinia pseudocasia, Populus alba, Celtis australis.

In order to sustain animal production. It is essential that the optimum feeding should be maintained throughout the year. During rainy season, green fodder is in surplus but no green fodder is available during May-June and Nov.-Dec. in these areas. It is, therefore, essential to conserve surplus forages to feed the animals adequately during these two lean periods and this can be done either in the form of silage or in the form of hay. Arya (1988 b) reported that very good hay/leaf meal can be prepared from *Robinia pseudocasia* loppings after chaffing and drying it in the sun. Similarly, good quality silage can be prepared from *Setaria*, hybrid *Napier* and grassland grasses in combination with tree leaves. According to Singh (1989) good quality silages are also obtained from natural grasses preferably at flowering stage after chaffing. In case dry matter is above 45 per cent, much attention is to be paid in pressing. Nutritional quality of grass silage (50% DM) can be further improved by mixing it with 1 per cent urea (on fresh weight basis sprayed with 1:1 solution in water) at the time of ensiling. In another study, Singh (1989) prepared silage from *Leucaena leucocephala* chaffed leaves with 30-60 per cent dry matter. It can also be ensiled with chaffed *Cenchrus ciliaris* grass in 1:1 ratio. Similarly good quality silage can be prepared from *Stylosanthes* with or without grass (1:1 ratio) after chaffing and harvesting at flowering stage. Stocking of forages sometimes reduces the toxic constituents present in the original forages for example, mimosine in *Leucaena* is reduced to about half due to ensilage.

Silvipastoral technology also helps in improving the palatability of poor quality forages and the palatability of local grasses and popular leaf fodder can be increased by mixing these fodders with *Robinia* leaf fodder and/or treating with salt, molasses or salt molasses additives (Arya, 1988 b).

REFERENCES

Arya, S.P. (1988a). Studies on the leaf fodder, fibre and fuelwood yields from *Grewia optiva* in relation to its girth class. Annual Research Report 1988-89: 37-38. Deptt. of S.A.F. UH&F, Solan.

Arya, S.P. (1988b). Studies on the improvement of tree leaf fodder. Annual Research Report 1988-89: 35-37. Deptt. of S.A.F., UH&F. Solan.

Arya, S.P. (1989c). Substitution of concentrate by *Grewia optiva* leaves in the ration of lactating cows. Annual Research Report 1988-89: 107-110. Deptt. of S.A.F., UH&F, Solan.

Arya, S.P. (1989d). Natural grassland production as influenced by tree species under silvipastoral system. Unpublished Data. Deptt. of S.A.F. UH&F, Solan.

Deb Roy, R., B.D. Patil, P.S. Pathak and S.K. Gupta (1980). Forage production of *cenchrus ciliaris* and *cenchrus setigerus* under silvipastoral system. *Indian J. Range Mgmt.*, 1:113-119.

Deb Roy, R. (1988). Biomass production potential of few fodder cum fuel trees under silvipastoral system. *National Workshop on research and extension needs for promotion of fodder and fuel trees,* Pune—July 4-7, 1988.

Deb Roy, R. (1989). Silvipasture production systems—present status and future prospects in India. *Lecture delivered at the Ist NARP Training Programme on Agroforestry, Forage Production and Animal Nutrition* at IGFRI, Jhansi Dec. 26, 1989-Jan. 25, 1990.

Deb Roy, R. and P.S. Pathak (1983). Silvipastoral research and development in India. *Indian Rev. Life Sci.* 3:247-264.

Hazra, C.R. (1989). Forages from degraded grasslands. Technology for increasing forage production in India. ICAR. New Delhi. pp. 1-10.

Saha, R.C. and B.N. Gupta (1987). Trees leaves as feed for dairy cattle in India. *Indian Dairy man,* 39(10): 489-492.

Sharma, S.K., K.S. Verma and V.S. Mishra (1988a). Silvipastoral studies in Northwest Himalaya. *Annual Research Report* 1988-89: 81-82. Deptt. of S.A.F. UH&F, Solan.

Sharma, S.K., K.S. Verma and V.K. Mishra (1988b). Natural grass yield as influenced by various fodder tree species under pastoral system. *Annual Research Report* 1988-89: 83-84 Deptt. of S.A.F., UH&F, Solan.

Sharma, S.K., K.S. Verma and V.K. Mishra (1988c). Leaf fodder production in relation to plant height and girth class. *Annual Research Report* 1988-89: 85-89 Deptt. of S.A.F., UH&F, Solan.

Singh, A.P. (1989). Conservation of fodder as silage and hay *Lecture delivered to the participants of Ist NARP Training on Agroforestry, Forage production and Animal nutrition at IGFRI,* Jhansi Dec. 26, 1989—Jan. 25, 1990.

Singh, P. (1989). Forage and fuelwood production through agroforestry systems in Wastelands. *Lecture delivered at the Ist NARP Training Programme on Agroforestry, Forage production and Animal nutrition at IGFRI*, Jhansi Dec. 26, 1989—Jan. 25, 1990.

Chapter 8

Agroforestry Research on Salt Affected Soils in India

*K.K. Mehta**

With the development of technologies for reclamation of salt affected soils, the importance of 'Agroforestry' research of salt affected soils has also been enhanced. Productivity of 952 million ha in the world in reduced by high soil salinity and alkalinity (Szabolcs 1977). These soils are spread mainly in the arid, semi-arid and in the coastal regions. Salt affected soils are mainly of two types i.e. saline and alkaline soils. Saline soils exists mostly in areas having less than 500 mm annum rainfall adversely affect the plant growth due to the presence of excessive soluble salts like chlorides and sulphates of sodium, calcium and magnesium. The electrical conductivity values of the saturated soil paste are generally more than 4 dSm^{-1}. Many crops and trees are however known to tolerate much higher concentrations of these salts. In case of alkali soils which occures in areas having more than 500 mm annual rainfall, the plant growth is adversely affected due to the presence of high exchangeable sodium caused by salts like carbonates and bicarbonates of sodium (EPS generally more than 50 in soil). Many plants like pea and gram suffer even at ESP levels of 15 whereas wheat tolerates up to 30 ESP and rice grows well even at 50 ESP. Generally the Ph (1:2) values range from 8.5 to 10.6. Infiltration rate and hydraulic conductivity values are very low in this type of soil. There is about 2 to 4 per cent calcium carbonate in the surface soil which increases with depth. At many places a 'Kankar' layer is found at about one meter deep. The presence of Calcium carbonate in these soils helps during reclamation as it releases calcium on solublisation which replaces excessive sodium from exchange complex.

In many developed countries like USA where population is less and food grain production is in excess to their requirements the need for

**Central Soil Salinity Research Institute, Karnal-132 001 (Haryana.)*

utilisation of salt affected soils for crop production and tree plantation is lesser than in developing countries with high population. Fertile lands near cities and towns are going out of cultivation as these are needed for housing and factories. The demand for food grains, fodder, fuelwood and timber is increasing every year. Under such situations it is essential to utilise even the salt affected soils for cultivation or tree plantation so that it is economically beneficial to the farmer and helps to improve the natural environment.

In India more than seven million ha of land was salt affected in early seventees (Table 8.1). With the establishment of Central Soil Salinity Research Institute,Karnal in 1969 and Agricultural Universities in different

Table 8.1 : Salt Affected Soils in India

Sr. No.	Broad group	Affected States	Area (000 ha)
1.	Sodic soils of the Indo-Gangetic plains.	Uttar Pradesh, Haryana, Punjab, Bihar, Madhya Pradesh and Rajasthan	2500
2.	Salt affected soils of the arid and semi-arid regions	Punjab, Haryana and Uttar Pradesh	1000
3.	Salt affected soils of the medium and deep black soil regions.	Karnataka, Madhya Pradesh and Maharashtra	1420
4.	Coastal salt affected soils		
	(a) Coastal salt affected soils of arid regions	Gujarat	714
	(b) Deltaic coastal salt affected soils of the humid region	West Bengal, Orissa, Andhra Pradesh Tamil Nadu.	1394
		Total :	7028

states, systematic research on reclamation of salt affected soils was undertaken. As a result, a tentative technology package for reclamation of sodic soils using gypsum was evolved at the CSSRI, Karnal. This technology was tested and improved on farmers fields so it became practically acceptable and economically viable (Mehta 1974, Mehta and

Abrol 1975, Mehta *et. al.*, 1975, Mehta 1989). In addition to gypsum, other amendments like pyrite and phosphogypsum were found useful for reclamation of alkali soils (Mehta and Abrol 1976; Mehta and Yadav 1977). Reclamation technology for crop production is briefly given below :-

1. Level and bund the alkali land. Make strong boundary bunds to prevent entry of water from outside.
2. Install a tubewell for irrigation even if canal water is available. In case the water table is high, emphasis should be more on tubewell irrigation as it will help to lower down the water table.
3. Grow rice at the 1st crop after application of gypsum or pyrite. High yielding varieties like IR-8, Jaya, PR-106, PR-103 and CSR-10 respond well. Nursery should be in normal soil. Once good crops are grown on alkali soil for one year, nursery can be in the reclaimed field as well.
4. Powdered gypsum passed through 2mm sieve should be applied (Broadcast). Mix gypsum in the upper 8-10 cm soil. One hectare of land should be divided into 5-6 sub plots. Irrigate and keep the water standing for 8-10 days. However, transplanting can also be done even after 2-3 days of irrigation. 30 to 35 days old healthy seedlings grown on a normal soil should be transplanted. At the time of transplanting the level of standing water should be less (4-5 cms). No puddling of soil is required before transplanting for initial two years.
5. Among fertilizers, apply a total of 150 kg N and 25 kg zinc sulphate per ha. Nitrogen should be applied in four splits i.e. 25 per cent basal, 25 per cent 20 days after transplanting, 25 per cent 30 days after transplanting and 25 per cent of prepenicle initiation stage. In the 2nd year, nitrogen should be 125 kg/ha and zinc sulphate 12 kg/ha. After 4-5 years 40 kg P_2O_5/ha should also be applied. Application of Potash should be done based on soil tests.

Different approaches to utilise alkali soils

There can be following approaches to utilise alkali soils:

1. Reclaim alkali soil by application of gypsum, phosphogypsum or pyrite and then grow rice-wheat in the initial few years as

described above. Other crops like berseem, jowar, mustard, mung, sunflower, potato, onion etc. can be grown as the soil improves within 4-5 years.

2. Reclaim alkali soils for crop production and then grow trees like eucalyptus, poplar on fruit trees for better utilisation of water and labour. Crops like berseem, can be grown in between the trees during initial 3-4 years (Mehta, 1989a).

3. Some salt tolerant short rotation trees like Valaiti Babool (*Prosopis juliflora*), Kikar, (*Acacia nilotica*), Eucalyptus may be grown directly on alkali soil by spot treatment i.e. by making pits or auger holes. Some of the fruit trees like Aonla (*Emblica officinalis*), can also be grown (Singh, 1989a). In between rows of trees, tolerant grasses like Karnal grass, can be grown (Singh and Gill 1990).

The adoption of any one of the above approaches depends upon the users. The users can be (1) Farmers. (2) Land Reclamation Corporations. (3) Forest Deptt. and (4) Panchayats.

Past experiences have shown that among the users only individual farmer made profit after initial expenses incurred by them in soil amendments, while both Land Reclamation Board and the Forest Department incurred losses in their attempt to utilise these lands after appropriate soil amendments. The last two being government departments. failures can be attributed to inadequate manpower and investment on account of which critical farm operations were not carried out in time, and lack of interest and motivation in timely maintenance of farm equipment and machineries during critical periods of farm operation. These were also compounded by corruption on the part of executors.

Crop cultivation

The technology evolved at CSSRI, Karnal and other Institutes includes land levelling, soil amendments with gypsum/pyrites etc. and then growing crops. The profitable crop rotation is rice-wheat. By following different steps, the farmers in seven villages (Kachhwa, Sagga, Sambhli, Bir-naraina, Guddha, Begampur and Dadlana) under Operational Research Project in Karnal district were able to get 45 Q/ha rice on demonstration plots set up in highly deteriorated alkali soils having pH values more than 10.2 and EC (1:2) values of 2.6 dSm^{-1} (Mehta & Mondal, 1988). Even in the initial years they were able to get 30 Q/ha of rice in

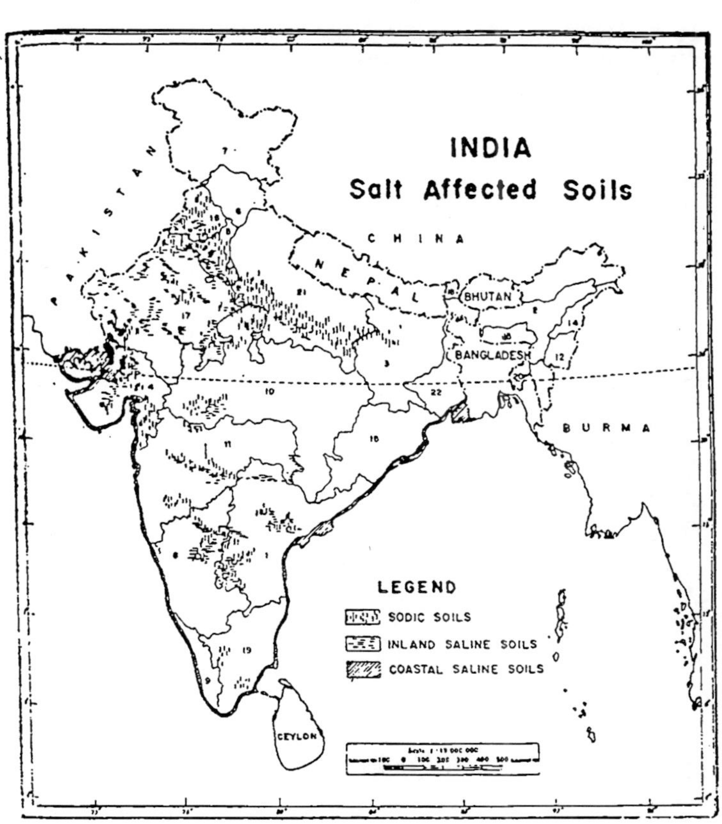

Fig 8.1 : Map showing saline and alkali soils in India.

the amended areas other than the demonstration plots. Farmers were able to get 12 to 20 Q/ha of wheat also in the Ist year after reclamation. The yields progressively increased due to continuous improvement of the soil brought about by cropping and irrigation. A survey conducted on 20 farmers fields revealed that rice yield increased steadily and got stabilised at about 60 Q/ha. Wheat yields stabilised about 30 Q/ha (Fig. 8.2). There is still possibilities of wheat yield increase.

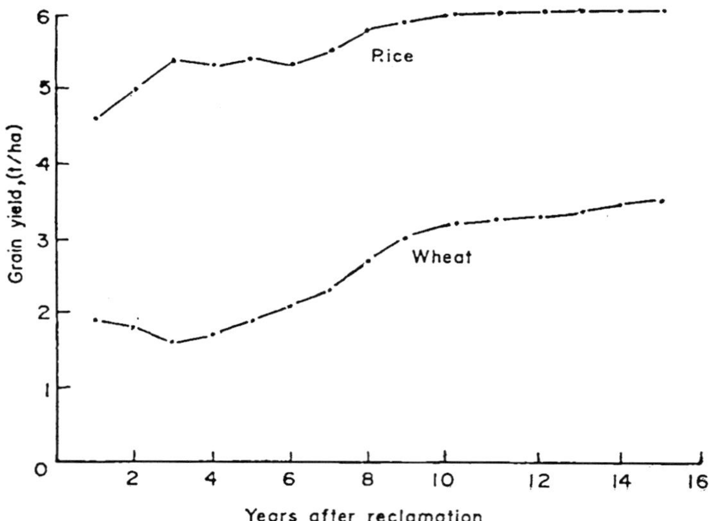

Fig. 8.2 : Grain yields of Rice and Wheat on demonstration plots on reclamation alkali soils in Karnal district.

The costs of reclamation were recovered within 2 to 4 years and an additional employment upto 140 man days per ha per annum was generated (Mehta & Mondal, 1988). All the alkali affected lands in the seven ORP villages were reclaimed by the farmers at their own cost. These lands (2500 ha) are now producing additional 20,000 tonnes of rice and wheat per annum where nothing could grow earlier. It has brought prosperity to poor farmers as additional gross income of Rs. 400 million accrue from these lands every year. It has helped the land less labourers also who get additional employment of 350,000 man days per annum. ORP villages in Karnal district is now treated as the model areas and learning spots, thus accelerating reclamation work on a large scale in Haryana, Punjab, Uttar Pradesh and Rajasthan. These studies have proved that alkali soils can be reclaimed economically and these should not be treated as waste or marginal lands.

Tree Plantation

Eucalyptus Plantation on Reclaimed Alkali Soils

Once the alkali soils were reclaimed, the farmers in ORP villages were encouraged to plant trees like eucalyptus near tubewells, on field bunds and even in blocks. Due to higher anticipated economic returns and its erect growth pattern the farmers took to eucalyptus planting on a large scale. Although the trees grew very well on reclaimed alkali fields, the economic returns were less than anticipated. Major returns were received at the end of 9 to 10 years which is very long waiting period of farmers. Lack of marketing facilities was the principal cause of low financial gains by the farmers.

A case study

Here is a case study of S. Darshan Singh of village Kachhwa who had reclaimed his 4 ha of alkali land in 1975 by gypsum application. He got good crop yields and in 1981 planted 0.4 ha land under eucalyptus. During this period, soil surface pH was reduced to 8.7. A total of 600 trees were planted without any gypsum application. Periodic height and girth observations revealed satisfactory growth. During the initial three years berseem crop (*Trifolium alexandrinum*) was grown during winter in between the eucalyptus plantings. This was sold at Rs. 2500 per ha. In fourth year, earthing was done so as to provide support and avoid dislodging of trees during rains. This blocks plantation was sold outright for Rs. 35,000 or say Rs. 97,500 per ha in December, 1989 and the field was cleared in January 1990. At that time the rate of eucalyptus by weight was Rs. 35 per quintal. The field was cleared from roots as well by the purchaser. There was some income from the loppings and fallen trees in between also. Total details of income and expenses in eucalyptus are presented in Table 8.2. The main findings were :

1. The net economic returns from rice plus wheat crop were Rs. 3800 to 4400 per ha per annum. The total net returns were Rs. 35750 during 1981-90. After adding interest @ 10 per cent it amounted to Rs. 53,946. Thus the net returns per annum were Rs. 5472 per ha. On the other hand the net returns from eucalyptus were Rs. 4263 per ha per annum which were less than that of crop cultivation. However, the farmer invested Rs. 7050 ha in the first year and continued to invest Rs. 3000 to 4800 up to 8th year. It appears expenditure after 3rd year was incurred on burseem which was sold at a loss (Rs. 2500 ha). Further, no attempt was made to utilise coppice shoots after first harvesting as the purchaser dug out the root portion as well. These explain the apparent loss compared to crops (Editors).

2. The economic returns from crop production were regular i.e. after every six months, whereas in case of eucalyptus the returns were mainly at the end of nine years though two coppice crop at shorter rotation should have been taken (Eds.).

3. There were adverse effects of eucalyptus on the nearby crops after four years of growth.

Due to the above factors growing of eucalyptus has been generally discontinued by the farmers even on semi-reclaimed alkali soils. The situation is more unfavourable for growing trees directly on unreclaimed alkali lands as is evident from the following experiments :

Growing Trees Directly on Unreclaimed Alkali Lands

From time to time research was carried out on planting trees on unreclaimed alkali soils. Different methods/techniques adopted were. (1) Pit Method, (2) Auger hole method, (3) Ridge and trench method.

Table 8.3 : Sodicity and salinity tolerance of two year old** Prosopis juliflora **(Valaity Babool) studied at Sikanderpur in Hardoi district of Uttar Pradesh.

Condition of the Prosopis plants	No. of sites	Root zone soil Properties				Plant Growth	
		Depth (cm)	pH_2	EC (1:2)	ECe	Height cm	Girth at coller cm
I Died or Very Poor	10	0-15	10.1-10.5 (10.2)	1.5-10.2 (5.9)	6.0-40.8 (23.3)	10-45	0.9-2.0
		15-30	10.1-10.5 (10.3)	1.3-4.2 (3.1)	5.2-16.8 (12.2)	(23)	(1.2)
II Medium Growth	10	0-15	10.0-10.4 (10.2)	2.0-7.6 (4.3)	8.0-30.4 (17.2)	200-300	12.0-15.0
		15-30	10.1-10.4 (10.3)	1.5-3.9 (2.8)	6.0-15.6 (11.1)	(237)	(14.0)
III Good Growth	10	0-15	9.6-9.8 (9.3)	1.0-1.2 (1.1)	4.0-4.8 (4.4)	425-475	20.0-25.0
		15-30	9.7-9.9 (9.8)	0.7-0.9 (0.8)	2.4-3.5 (2.9)	(450)	(23.0)

Values in parenthesis show average.

Table 8.2 : Comparative net economic returns from crop cultivation and Eucalyptus planting on a semi-reclaimed alkali field in village Kachhwa during 1981-90.

Year	Grain yield Q/ha Rice	Wheat	Net returns Rs./ha	Expenses on Eucalyptus berseem growing (Rs./ha)	Loss due to Eucalyptus on adjacent crops.(Rs./ha)	Returns from sale of Eucalyptus & berseem Rs./ha
1981-82	59	27	3800	7050	-	2500
1982-83	58	29	3800	4700	-	2500
1983-84	57	30	3850	4700	-	2500
1984-85	60	30	4000	4900	-	-
1985-86	56	32	3800	4800	380	1000
1986-87	54	30	3900	4700	400	-
1987-88	57	30	4100	4000	550	-
1988-89	56	30	4300	3800	660	-
1989-90	55	31	4400	3000	960	87500
			35750	41650	2950	96000
Interest up to 1990			18196	23876	485	11326
Total			53946	65526	3435	107326
After deducting expenses & loss to crops						38365
Net returns Rs./ha/annum			5472			4263

Pit method

Yadav *et al.* (1972) planted trees by digging pits of 1 x 1 x 1m size with the following treatments : (i) the original soil replaced with normal soil, (ii) soils teated with gypsum and FYM. Treatment with gypsum + FYM was yield better results than soil replacement. The trees have been growing for more than twenty years. If these are sold it will be found very uneconomical in comparison to crop cultivation. This techniques could not become popular for growing trees as the initial investment were very high and there was no guarantee for satisfactory returns.

During 1986, 10 ha of salt affected land of Alisabad and Sikanderpur villages in U.P. was put mainly under Valaiti Babool (*Prosopis juliflora*) by the forest Department. of Mander district. Pits of 60 x 60 x 60 cms were dug and were refilled with the same soil after treatment with 5 kg pyrite and 10 kg sand. It was observed that during the initial 2-3 months the mortality was high and about 10000 plants were replaced. In February 1987 about 10000 additional dead plants were also replaced. Drying of the trees started from top and continued downwards. During May 1988 soil samples from the root zone of these trees were examined and growth was also measured. It was observed that on an average pH levels of more than 10.2 and ECe levels of 23.3 dSm^{-1} existed at 0-15 cm soil layer which might be responsible for poor growth or death (Table 8.3).

Good growth was observed at pH levels of 9.7 and ECe levels of 4.4 careful perusal of the data revealed that lowering higher pH levels in the root zone toxicity due to salt concentration were in the dominant factors for the poor growth/mortality of the trees.

Fruit trees in alkali soils

Fruitful research on raising of fruit trees on alkali soils has been carried out at Narendra Dev Agricultural University, Faijabad. The alkali soils had pH values around 9.5 at the soil surface. Aonla, Karonda, Ber, Jamun, Phalsa, Grape, Guava and Bael were planted by making pits of about 1 x 1 x 1m size (Pathak *et. al.* 1990). In addition to gypsum, lot of farm yard manure (FYM) was also applied in the pits (Table 8.4). Salt tolerance and productivity of different fruit trees at fruit bearing stage has revealed that Aonla was the most alkali tolerant among the eight fruit trees (Table 8.5). These trees were grown at the research farm under the constant guidance and supervision of scientists. The usefulness of these studies requires testing on the farmer's fields and the cost benefit ratio of

Table 8.4 : Details of pit preparation and planting of aonla, ber and bael

Fruit crops	Cultivars	Pit size (cubic m)	Soil-mixture				
			Soil	FYM (kg)	Sand (kg)	Gypsum/ pyrite (kg)	
Aonla	Chakaiya Kanchan Krishna	1.00	Top Soil	40	30	5	4
Ber	Gola Banarasi-Karaka Umran	0.75	Top soil	30	20	5	4
Bael	Faizabad Selection	0.75	Top	30	20	5	4

growing fruit trees in comparison to crop production has to be calculated before making any recommendation for large scale adoption of this technology.

Table 8.5 : Salt tolerance and productivity of some fruit plants at full bearing age

Fruit crops	Salt tolerance (pHe)	Age (years)	Productivity (t/ha)
Aonla	10.0	8 and above	20.5
Karonda	10.0	5 and above	5.2
Ber	9.5	5 and above	15.5
Jamun	9.5	6 and above	16.0
Phalsa	9.2	3 and above	6.0
Grape	9.0	4 and above	18.3
Guava	9.0	4 and above	12.5
Bael	8.5	8 and above	16.5

Auger hole method

During eightees research on use of Auger hole for planting trees on alkali soils at CSSRI research farm Guddha was started (Abrol and Sandhu 1980). The diameter of hole was mainly 15 cm and depth varied from 120 cm to 180 cm in different experiments. Following conclusions can be drawn :

(i) Breaking of 'Kankar' pan that exists at about 1.5 m depth is essential to grow trees.

(ii) 3 kg gypsum and 8 kg FYM mixed with the original alkali soil is a suitable filling mixture.

(iii) Root growth was very good only at the initial stages which was restricted to the auger hole. Plant growth was also good in the initial stages but later-on the rate of growth was retarded substantially. Among trees, growth of eucalyptus was adversely affected after 3 years.

(iv) *Prosopis juliflora* (Mesquite) or Valaiti Babool was the most successful. During a period of about 9 years the total biomass was about 270 tonnes/ha. Cost of growing valaiti babool was Rs. 10500 per ha (Singh *et. al.*, 1990). The author (Dr. K.K. Mehta) had experience of planting trees by augèr hole on farmers fields in operational Research Project villages. It was observed that making auger holes was very difficult and expensive if done manually during May and June. It was not at all possible to break 'Kankar' pan. The author also modified the calculations with regard to cost of auger hole from Re. 1 per hole to Rs. 3 per hole as this is the minimum cost when cost of Auger is included with tractor Augers get broken when thick layer of 'Kankar' pan is encountered. Nominal charges for maintenance and care of the plants was also included which was not included by Abrol and Singh. During 1st year the cost came to Rs. 16,000 per ha (Table 8.6) in which cost of making auger holes, application of gypsum, FYM and planting of saplings was included. Total cost of growing valaiti babool was Rs. 33000 over a period of nine years. It amounted to Rs. 58854 when interest @ 10 per cent was added to it. This cost is still on the lower side as lot of labour charges for growing these trees at the research farm were not included. There were returns of Rs. 500 to 1000 per ha due to loppings from 4th year onwards. At the end of nine years the expected returns were Rs. 75176 and the net gain was Rs. 16322 over nine years which came to Rs. 2040 per ha per annum. On the other hand the expected returns from eucalyptus were on the negative side with a net loss of Rs. 1244 per ha per annum (Table 8.6). These losses are the main reasons that most of the farmers do not grow eucalyptus or *Prosopis juliflora* on alkali lands. From the present studies it is crystal clear that alkali soils not being treated as either waste or marginal land for tree planting is not profitable and the current efforts by the State Forest Departments should be distinued in favour of utilising the land for more remunerative crops. Cost computation included average expenditure of Rs. 2000 per annum, per hectare from 4th to 9th year for after care. Except prevention of illicit felling, no aftercare is normally needed for forest trees. This has made calculation of cost benefit ratio somewhat fallacious.

Table 8.6 : Cost of growing Valaiti Babool (Prosopis juliflora) and Eucalyptus on a highly deteriorated alkali soil (pH 10.0) by auger hole method

Year	Particulars	Prosopis juliflora		Eucalyptus	
		Expenses	Returns	Expenses	Returns
1st	Auger holing, gypsum, planting	16000	—	16800	—
2nd	Irrigation and after care	3000	—	3000	—
3rd	After care	2000	—	2000	—
4th		2000	500	2000	—
5th		2000	500	2500	-
6th		2000	1000	2000	—
7th		2000	1000	2000	500
8th		2000	1000	2000	500
9th		2000	70000	20000	50000
		33000	74000	34300	51000
Interest @ 10 per cent		25854	1176	26761	105
		58854	75176	61061	51105
Gain/loss			16322		-9956
Rs./ha/annum			2040		-1244

Growing grasses in between *P. juliflora* (V. Babool) trees

Research work on growing of different grasses in the inter space between *Prosopics juliflora* trees was carried out at CSSRI research farm Guddha in Karnal district. The trees were planted mainly by auger hole technique. Growing of slat tolerant grasses like Karnal grass was possible and their yield cuttings were available mainly during rainy season. It is evident from the work reported by Singh and Gill (1990) which is presented in Table 8.7. The success was possible due to intensive care and inputs on weeding, irrigation etc. If all the expenses are included it will be very uneconomical but the problem is that all costs were not included and one does not get a true picture. Grasses like Karnal grass grow well on alkali soil but they are not palatable to cattles, therefore, difficult to market the produce.

Table 8.7 : Yield of Karnal grass grown in association with mesquite (Prosopis juliflora)

Planting year	No. of cuttings	Cuttings months	Forage yield t/ha
1st	1	November	2.2
2nd	4	May to October	13.1
3rd	3	July to September	10.1
4th	3	May to September	7.6
5th*	4	July to October	13.5
Total	15	-	46.5

*Karnal grass was ploughed under and other useful fodders were grown.

Whenever, calculations are made the actual sale price which is mostly very less is not reported which leads to different conclusions. Lopping of valaiti babool helped in better growth of grass as reported by Valayati babool helped in better growth of grass as reported by Sing and Gill (1990). However, owing to the presence of dangerous thorns, lopping is difficult and problematic. Moreover, the actual cost involved in these loppings were not reported. Such missing items in the cost computation leads to erroneous conclusions.

Effects of trees on nearby crops

Eucalyptus, Babool (*Acacia*) and poplar were the main trees species which were grown on semi-reclaimed alkali soils on a large scale. Eucalyptus was planted either in blocks or on the boundary and field bunds. Babool was planted on the road sides in areas where alkali soils existed. Eucalyptus seem to cause adverse effects on nearby crops, other two tree species did not cause any apparent adverse effects.

Eucalyptus

Studies conducted in village Kachhwa on the effects of Eucalyptus plantation (planted in 1981) on the nearby rice and wheat crops led to the following conclusion (Mehta & Mondal, 1988).

1. During the initial four years of eucalyptus growth very little or no adverse effects on the growth and yield of nearby rice crop was experienced (Table 8.8).

Table 8.8 : Grain yield and growth of rice (PR-106) as affected by four year old Eucalyptus trees planted on partially reclaimed alkali field bunds.

Distance from Eucalyptus row (meters)	Average height at maturity (cms)	Av. number of tillers per spot	Av. Grain yield kg. per 4 m	Grain yield Q/ha
0 - 2	79.6	10.2	2.20	55.0
2 - 4	84.8	11.3	2.50	62.5
4 - 6	91.4	13.6	2.60	65.0
6 - 8	88.0	12.2	2.70	67.5
8 - 10	86.2	13.6	2.90	72.5
10 - 12	82.2	12.1	2.75	68.7
12 - 14	89.0	12.0	2.50	62.5
14 - 16	86.2	12.5	2.60	65.0
16 - 18	86.2	12.5	2.60	65.0
18 - 20	84.2	12.0	2.90	72.5

2. Wheat yields decreased both near the block plantation or row plantation of Eucalyptus which is attributable to : (a) decrease in soil moisture content near the trees, (b) Eucalyptus shade causing delay in showing of wheat due primarily to continued moist condition of soil on the northern side of the eucalyptus rows which were planted in East-West direction.

3. No 'allelopathic' effect of eucalyptus was observed on plants of other species that were growing as "off plants" in the block plantation of eucalyptus.

4. There was no significant improvement on pH values at different distances from the eucalyptus row. However, there was trend of reduced pH near the row relative to the distant places.

Avoiding Adverse Effects of Eucalyptus

As the adverse effects were due to shade, moisture and nutrient depletion following remedial measures are suggested (Mehta and Mondal, 1988).

1. Eucalyptus planting on field bunds should be in North-South direction so as to minimise shade effect.

2. Water channel should be provided along the trees rows for trees to get sufficient moisture.

3. Additional dose of fertilizers should be applied near the tree rows so that crops do not suffer due to competition from the trees.
4. The distance between tree to tree should be more than two meters.

Agroforestry on saline soils

In India, systematic research on reclamation of saline soils has been carried out only in the recent years and accordingly very limited work on agroforestry has been done. Saline occur in more than 4 million ha of land mainly in the arid and semi arid areas in the states of Gujarat, Rajasthan, Andhra Pradesh and Bihar, Haryana, Punjab, Karnataka, Maharashtra, Madhya Pradesh and Uttar Pradesh (Table 8.1). Chemical analyses of two soil profiles from Bhal area of Gujarat (Table 8.9) revealed that these soils have high salt contents in the entire soil profile and soils are heavier in texture with high clay content. Further, groundwaters are mostly brackish which are not suitable for irrigation and canal irrigation often leads to secondary salinisation. This happens mainly due to excessive use of canal water at the "head end" especially when canal distributories are not completed up to the tail end. This leads to rise in ground water table which in turn caused waterlogging and salinity problem in some pockets. Before reclamation of saline soil, the extent of salinity has to be measured and soil properties of the whole profile have to be studied. Total rainfall and its distribution; water table depths and their fluctuations etc. are other important studies. The problem of soil salinity in arid and semi-arid regions is so varied that different combinations of amendment measures have to be adopted which may be as follows :-

(1) *Problem*: Soil saline, water table deep and canal water available.

Solution : Leaching will help to push the salts downward. Crops requiring excessive irrigation like sugarcane and rice should not be grown as it can raise water table and create further waterlogging and salinity problem.

(2) *Problem* : Soil saline, water table deep and ground water brackish and canal water absent.

Solution : Salt tolerant trees, bushes or crops should be grown keeping in view the amount and distribution of rainfall. Some of the important tree species tolerant to salinity are *P. Juliflora, Acacia nilotica, Tamarix* spp. *Casuarina equisetifolia* and *Eucalyptus camaldulensis.*

Table 8.9 : Physiochemical and morphological characteristics of soils of Bhal

Horizon	Depth (cm)	pHe	ECe (milimhos/cm)	ESP	Organic C (%)	$CaCO_3$ (%)	Clay (%)	Infiltration rate after 6 hrs (cm/hr)
1	2	3	4	5	6	7	8	9
Coastal area below 5m contour : Sanes, Vertic Halaquepts								
A_1	0-13	8.0	34.0	50.5	0.01	14.1	38.3	0.25
B_1	13-34	8.2	38.0	60.3	0.05	18.2	39.9	0.27
$B_{21}Sa$	34-62	8.2	89.2	63.8	0.03	13.8	41.2	0.41
$B_{22}Sa$	62-109	8.3	110.0	70.9	0.04	17.1	44.2	0.67
C	109-155	8.4	39.5	87.6	0.01	20.0	38.8	0.59
Savaikot, Typic Halaquepts								
C_1Sa	0-4	8.5	150.0	35.0	0.35	8.0	50.1	0.35
C_2Sa	4-50	8.6	170.0	40.8	0.30	8.9	50.0	0.30
C_3Sa-cs	50-180	8.6	180.0	55.9	0.38	9.5	35.0	0.25
C_4Sa	180-205	8.8	200.0	65.0	0.25	10.0	25.0	0.18
Area under canal irrigation commend : Varasada, Aquic Natrudalfs								
A_{sg1}	0-14	8.6	34.3	59.7	0.85	12.5	40.5	0.25
B_2t	14-42	8.8	28.2	61.8	0.65	16.0	50.3	0.15
C_1	42-78	8.9	25.1	67.4	0.61	8.0	50.1	0.10
C_{2g}	78-130	8.4	14.2	67.9	0.41	7.5	46.1	0.10

(3) *Problem :* Soil saline and waterlogged. Water table fluctuating within 0 to 3 meters. Canal water available in plenty.

Solution : Salt tolerant and waterlogging crops like rice and sugarcane may be grown. Check up ground water and if found suitable, use it though tubewell which will lowerdown the water table to some extent. Open or pipe drains may be required to drain out the excessive salts and water but due consideration should be accorded to the problem of disposal of drained out water and socio-economic aspects. If disposal problem is there the area may be put under trees like *Prosopis, Acacia, Eucalyptus* etc., depending upon severity of the salinity problem.

(4) *Problem :* Soil saline and waterlogged. Canal water available only in small quantity.

Solution : Grow salt tolerant trees shrubs by using limited quantity of canal water for establishing the trees/shrubs. For this, sub-surface method of planting (Fig. 8.3) may be adopted with advantage. The pit is kept 40 cm dia at the top and 30 cm dia at bottom and 20 cm deep. Planting at the end of August or early September are more suitable. Frequent irrigation helps in higher in survival and better growth.

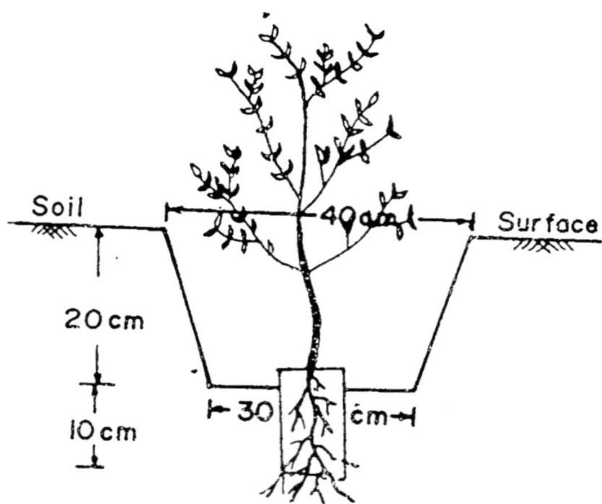

Fig. 8.3 : Subsurface method of planting tree saplings in saline soils

Crop Cultivation after Drainage

Systematic work on drainage and crop cultivation on saline soils was carried out at CSSRI Research farm sample in Rohtak district. The soils are medium textured alluvial soils and water table was high and fluctuated from zero to 2.5m approx. The soils became saline due to secondary salinisation. The EC of soil at many spots was around 100 dSm^{-1} in the saturation extract. That area of about 10 ha belonged to village Panchayat and was almost abandoned due to high salinity and waterlogging. The ground waters were brackish. These lands were reclaimed mainly under the guidance of Dr. K.V.G.K. Rao, Project Manager of the Indo-Dutch programme on reclaimation of saline soils. Underground pipe drains were laid at 1.5m depth at a spacing of 25, 50 and 75m and it was found that 50m spacing among drain lines was optimum for satisfactory crop production. Successful crops of wheat, mustard, pearl millet, cotton and barley have been grown on the once highly saline soils. The salt content of the soil also decreased due probably to cropping.

Inspite of the above success, several problems like availability of good quality canal water, cost of the project, maintenance of the system, clogging of drains, disposal of the drained out water etc. In the case under study drainage system was possible because good quality canal water was made available for this land as a special case by the irrigation department. Providing irrigation facility to the 4 million ha affected by salinity is a great problem. Another major problem is disposal of drained out water which is highly saline. Evaporation tank has been dug at Sampla farm but the cost involved is prohibitive. Moreover, the same water can find ways to the adjacent area from evaporation tanks and may affect the other piece of land. The cost of laying drainage system at 50m spacing between the laterals came to Rs. 14,000 approx. (Joshi *et. al.*, 1985). Cost of Rs. 14,000 per ha for 50 m spacing will require about Rs. 5600 million for laying out drainage system in 4 million ha of affected land which is a huge sum. Maintenance of piped drains is a constant problem and may prove more expensive than expected. For example it was expected that the pipes would last for 30 years and the economics of the project was calculated accordingly. Some of the pipes were examined after four years of working and it was observed that due to silting the pipes were clogged by 33 to 75 per cent (Gupta and Swaroop, 1990).

This is the situation when the drainage pipes were laid under the strict supervision of the Engineers. It will be very difficult to expect quality work from the contractors and the working of the drains and their life will be much less than expected. Flushing of the pipe drains will be another expensive item which was not included while making calculations. The huge cost of consultants from foreign countries will be an additional burden and it will drain out our precious foreign exchange.

If collectors or lateral drains are kept as open drains then weed problem is so excessive that it is very difficult and expensive to control it. It was observed by the author in USSR (Mehta, 1980). In USA also the drainage system laid out at huge expenses in Sanvaquine Valley in California State was closed due to pollution by the toxic elements in drained water (Mehta, 1989b). Although presence of toxic elements in our drained out water may not be high, the disposal of drained water will pose constant problem for reasons discussed earlier.

So it is advisable to learn lessons on drainage systems from USSR, USA, Egypt and Iraq and should be highly reticent in collaboration with other countries in getting the already known 'know-how' at exorbitant cost.

Role of trees in reclaiming saline soils - a case study

Trees have great potential in reclaiming saline soils which was evident from the studies conducted at CSSRI Research Farm Sampla during 1987-88. *Acacia nilotica* (Babool) were planted on the saline soil in 1984. During 1987-88 soil samples from 0-105 cm soil depth at 15 cm interval were taken during different months both from the *Acacia* field as well as from the adjacent barren field where no trees were planted. It was observed that -

1. Salt concentration was less in all the layers in *Acacia* planted field in compared to the barren saline field. At the soil surface (0-15 cm) the salt concentration was reduced to half or 1/4th (Fig. 8.4). During March it was 18.5 dSm^{-1} in barren field whereas it was only 4.8 dSm^{-1} in Acacia field. Similar was the trend in other months also.

2. Moisture content was almost static in surface as well as deeper layers in case of barren field in all the months. It was lower in the upper 60 cm soil layer in *Acacia* field especially during the summer months of May and June (Table 8.10) which seems attributable to combined effect of evapo-transpiration.

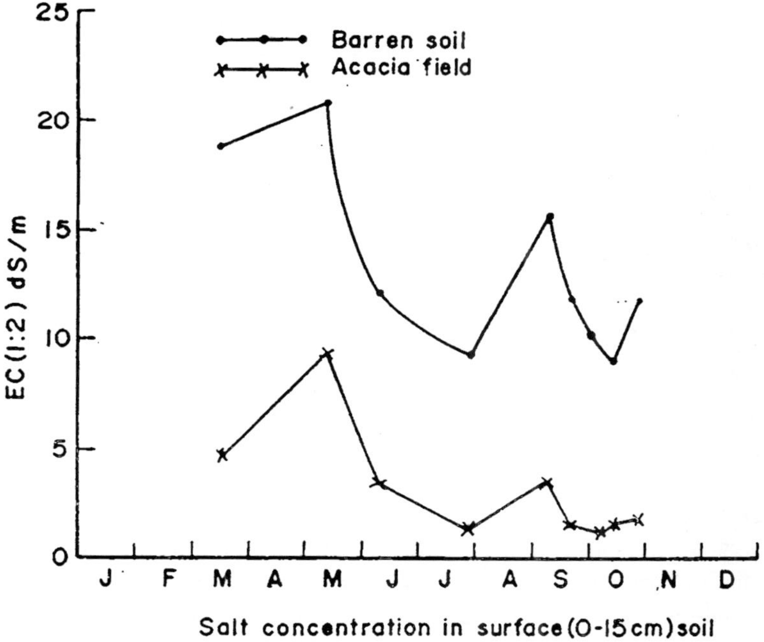

Fig. 8.4 : Effect of ***Acacia nilotica*** *on salt concentration under saline soils*

Table 8.10 : Moisture content (Per cent) of soil profile in barren and in adjacent babool (Acacia nilotica) planted saline fields at CSSRI farm Sampla, Rohtak district of Haryana

Depth	18.3.88		12.5.88		10.6.88	
	Barren field	Acacia field	Barren field	Acacia field	Barren field	Acacia field
0-15	13.25	13.64	12.00	5.12	9.1	3.8
15-30	13.25	14.29	11.34	7.58	9.5	7.0
30-45	13.77	15.07	11.67	10.42	11.2	7.9
45-60	14.42	16.01	13.57	12.17	12.7	9.0
60-75	16.55	16.01	14.02	12.72	12.4	11.1
75-90	15.74	16.41	14.29	12.88	13.3	12.3
90-105	15.34	18.20	15.29	13.07	14.5	13.0

During November, 1988, soil samples were also taken from the tile drained area, barren field and *Acacia* planted field at CSSRI farm Sampla. The salt concentration (Fig. 8.5) was maximum in barren field whereas it was minimum in tile drained field. Four years old *Acacia* also helped in reducing salt concentration to a great extent in the soil profile.

From the above data it is also evident that *Acacia* tolerated salt concentration of 9.6 dSm^{-1} in 1:2 soil : water suspension which corresponded to 38.4 dSm^{-1} in saturation extract in the month of May. Keeping in view these observations it can be concluded that under these agro-climatic conditions it may not be necessary to drain out such areas where trees can be planted successfully and these trees can use this water for producing fuelwood and timber. Moreover it is difficult to dispose of the drained out water due to pollution problems as explained earlier. For example within one year, the drained water seeped into the nearby village pond, the animals did not drink the polluted water and there was problem to deal with the angry villagers. Similarly the drain water cannot be put into canals as it will also pollute it which is also used for drinking by human being.

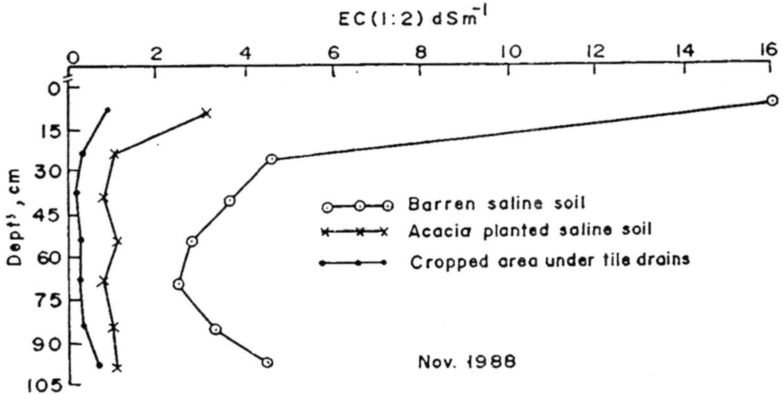

Fig. 8.5: Comparative salt concentration in Barren, Acacia planted and tile drained saline fields

For successful establishment of trees in such areas it is very essential to study and understand the water table fluctuations and salt and moisture dynamics so that a suitable period is selected for tree planting when salt concentration is the lowest at surface layers. Mehta (1988) observed that root zone salinity was lower after monsoons, making August/September ideal for tree planting. Sub-surface planting in the small pit of 20 cm deep (Fig. 8.5) promotes higher survival rate. The soil salinity could be avoided to another 10 cm depth by not removing the polythene packing from the sides. Only the bottom of the polythene bag in which sapling is lodged is removed. The upper 30 cm of soil salinity was thus avoided. Keeping higher moisture content at the root zone by frequent irrigations helped in better survival rate and growth of *Acacia* (Fig. 8.6). Watering could be done manually using a bucket. Higher moisture content near the root zone was also achieved by pitcher method of irrigation in which water was put in the pitcher, from where water oozed out and the roots with had constant moisture.

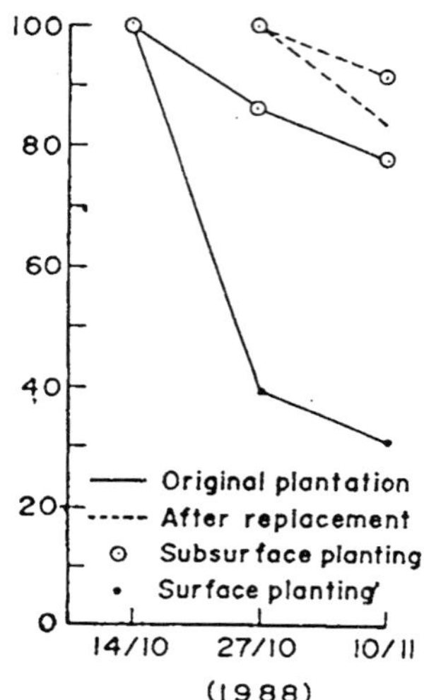

Fig. 8.6: Survival percentage by using surface and subsurface method of planting

Although more studies are needed on the role of moisture content and salt tolerance of *Acacia* and other trees but it was clearly evident that trees tolerated the salt concentration more where moisture content was high. For example growth was much better near water channels than away from them.

REFERENCES

Abrol, I.P. and Sandhu, S.S. (1990). Growing trees in alkali soils. *Indian Fmg.*, 30(6), pp. 19-20.

Gupta, S.K. and Swarup Anand (1990). Clogging of pipe drains in saline soils. *Ann. Rep.*, CSSRI, Karnal, pp. 90-91.

Joshi, P.K., Agnihotri, A.K. and Singh, O.P. (1985). Cost of sub-surface horizontal drainage for reclamation of saline soil. CSSRI, *Report (Econ)*, No. 1-5 (Pt. 1), p. 51.

Mehta, K.K. (1974). Grow bumper crops in alkali soils, *CSSRI information Bull.*, p. 8.

Mehta, K.K. (1980). Reclamation and improvement of irrigated saline soils. *Report on Training courses in USSR during 1978* : CSSRI, Karnal Report, p. 29.

Mehta, K.K. (1983). Reclamation of alkali soils in India. *Oxford andIBH Publishers* 66, Janpath N. Delhi p. 283.

Mehta, K.K. (1984). Eucalyptus plantings on semi-reclaimed alkali soils and its effect on nearby crops. *Ind. Soc. Soil Sci., Convention* (49th). Abstract of papers, p. 146.

Mehta, K.K. (1986). Prosperity through wasteland reclamation *Wastelands News,* 2(4), pp. 2-3.

Mehta, K.K. (1987). Basmata brings new hope for wastelands. *Wastelands News,* 2(4), pp. 5-6.

Mehta, K.K. (1989 a). Grow trees on waste alkali soils. *Indian Farming,* 39(4), pp. 25-28.

Mehta, K.K. (1989 b). Use of Poor quality water for agriculture. *Report on training in USA,* CSSRI, Karnal, p. 53.

Mehta, K.K. (1990). Experience of growing crop and trees on salt affected soils by Hardoi district in Uttar Pradesh in Agroforestry, (Singh *et. al.*, eds.), pp. 116-124.

Mehta, K.K. and Abrol, I.P. (1975). Reclamation of alkali soils-impact on Sangrur farmers. *Indian Fmg.*, 25(1), pp. 7-8.

Mehta, K.K. and Abrol., I.P. (1976). Pyrites for reclamation of alkali soils. *Indian Fgm.*, 26(1) pp. 20-21.

Mehta, K.K. and Yadav, J.S.P. (1977). Phosphogypsum for reclamation of alkali soils. *Indian J. Agric. Sci.* 47, pp. 348-54.

Mehta, K.K. and Mondal, R.C. (1988). Operational Research on Reclamation of alkali soils, *Bull. No.* 12, CSSRI, Karnal, pp. 66.

Mehta, K.K., Yadav, J.S.P. and Abrol, I.P. (1975). Reclamation of alkali soils. *CSSRI Inform. Bull,* pp. 2-16.

Pathak, R.K., Dikshit, S.N. and Dwivedi, R. (1990). Prospects of aonla, ber and bael cultivation on salt affected wastelands in India, In : *Agroforestry,* (Singh *et. al.* eds.), pp. 125-132.

Sandhu, S.S. and Abrol, I.P. (1981). Growth responses of *Eucalyptus tereticornis* and *Acacia nilotica* to selected cultural treatment in a highly sodic soil. *Ind. J. Agric. Sci.*, 51(6), pp. 437-43.

Singh, G. and Gill, H.S. (1990). Raising trees in alkali Soils. *Wastelands News*, Nov. 1990-Jan. 1991, pp. 15-19.

Yadav, J.S.P., Bhumbla, D.R. and Sharma, O.P. (1972). Performance of certain forest species on a saline sodic soil. *Proc. Symp.*, New Development in the field of salt affected soils. Int. Soc. Soil Sci. Conference Cairo, pp. 683-90.

Chapter 9

Present Status and Future Prospects of Integrating Agricultural Crops and Fodder in Temperate Himalayas

Dinesh Kumar* and S.D. Bhardwaj**

There are nearly 275 million large ruminants and another 165 million small ruminants in India. A large proportion of these are maintained as small herds by poor landless and marginal farmers for subsistence income. The present livestock population requires an estimated 343 million tonne green fodder and 347 million tonne dry fodder, apart from 20 million tonne concentrates. However, the present production is only 227 million tonne green fodder and 207 million tonne dry fodder, thus leaving an overall deficit of 36 per cent in the roughage requirements (Mudgal and Pradhan, 1988). The situation is likely to worsen in the years to come owing to the ensuing reduction in cultivable land caused by expanding construction activities and the increasing fodder requirements of our growing livestock population.

Green fodder crops occupy a mere 0.5 per cent of the total agricultural area in the western Himalayas. The grasslands constitute another alternative but these are few and generally far away from habitation. With the drying of grasses towards the onset of winter, the axe falls on trees. Tree foliage provides an excellent and nutritious fodder. Nomadic tribes move to the alpine pastures during summer for migratory grazing, this plays havoc with the ecosystem. The last 15 years have witnessed a rise in milk production by 19.3 per cent only, despite a phenomenal increase of 41 per cent in cattle population. Most of the forest areas have been subjected to over-grazing and severe lopping for leaf fodder. The situation on fodder front is so serious that many people set fire to dry grass and pine needle

**Department of Silviculture and Agroforestry*
***Dr. Y.S. Parmar University of Horticulture & Forestry Nauni, Solan - 173 230 (H.P.)*

The grasslands are dominated by *Panicum maximum, Chloris gayana, Chrysopogon montanus, Heteropogon contortus, Poa pratensis, Festuca pratensis, Setaria, Eragrostis curvula, Apluda mutica,* etc. interspersed with shrubs and trees of *Albizia chinensis, Grewia optiva, Celtis ustralis, Toona ciliata, Pinus roxburghii, Pistacia integerrima, Pyrus pashia, Punica granatum,* etc. In the natural grasslands under silvi-pastoral system, grass yield was found to be 7660 kg/ha in association with *Robinia,* 6350 kg/ha with *Grewia,* 5250 kg/ha with *Bauhinia* and 6730 kg/ha with *Albizia lebbek* (Anonymous, 1989a).

The palatability of local grasses and poplar leaf fodder is increased by mixing these fodders with *Robinia* leaves or on treatment with salt, molasses or salt plus molasses. In an experiment, poplar leaves as such were not eaten by the livestock. On treatment with 1 per cent salt plus 3 per cent molasses, 100 per cent intake of poplar leaves was observed (Anonymous, 1989b). Pine needle litter and shade have adverse effect on the growth of grass; these account for 30-50 per cent reduction in grass biomass. The removal of pine needle litter, however, restores grass yield to almost the same level as in the open area without pine tree (Anonymous, 1991).

Evaluation of horti-silviculture (plum + poplar) and silvi-pastoral systems is also underway at Solan. The Himachal Pradesh State Council for Science and Technology has desired to extend these studies to other zones of Himachal Pradesh (Anonymous, 1990).

More than 53 per cent of the total annual requirement of grass for livestock consumption is met from "ghasnies" owned by the farmers and village panchayats, while more than 51 per cent of the leaf fodder is obtained from forest areas and the rest from the privately owned trees (Swarup and Tewari, 1988).

(b) High-hill moist temperate zone (above 1600 m)

Wheat, barley, minor millets, buckwheat, amaranthus, French bean, black gram and maize are the common field crops. The mainstay of the farming community is fast changing from agriculture to horticulture. Apple, pear, potato, pea, cauliflower and other temperate vegetables are the major cash crops. Potato, chilly and sweet pepper are often raised in the apple orchards.

The farmers are dependent on the nearby forests for fodder and fuel. Some needs are met from the trees growing on terrace risers. *Aesculus,*

Alnus, Betula, Celtis, Juglans, Grewia, Morus, Populus, Quercus, Robinia, Salix, etc. are the major fodder trees. Forest trees are sometimes grown as windbreaks, shelterbelts or fillers in the orchard for protecting it from high-velocity winds. *Populs deltoides, Populus nigra* and *Salix* have been successfully tried around the orchard without any adverse effect on fruit production (Shamet and Puri, 1986).

Gupta (1983) opined that big trees reduce the yield of agricultural crops by 40 to 70 per cent while small trees do not affect production. *Salix, Grewia, Morus* and *Populus* generally have no adverse influence on agricultural production. In fact, soil and water losses are less from multi-tier system than single-tier agricultural crop.

Agricultural crops were raised between 5-7 years old *Populus deltoides* trees planted at 5m x 3m, 5m x 4m and 5m x 5m spacings.

The yield of all crops except raya declined in combination with poplar. The decline was more at closer spacing of poplar (Anonymous, 1987, 1988, 1989a, b).

Table 9.1 : Yield of crops (kg/hectare) in field planted with poplar

Poplar spacing	Wheat	Lentil	Gram	Soybean	Cowpea	Sun flower	Mus-tard	Raya
No poplar	623	556	440	722	613	540	200	120
5m x 3m	348	148	128	183	105	100	87	97
5m x 4m	355	122	346	191	123	119	112	122
5m x 5m	493	168	364	223	112	144	167	259

Ref. Anonymous (1987, 1988, 1989a, b)

In another experiment involving poplar trees at 5m x 3m spacing, grain yield of wheat var. S-308 was maximum (41.91 q/ha) when wheat was sown on November 15 and applied 150 per cent of the normally recommended doses of nitrogen without using poplar leaf mulch. When sowing date was advanced, grain yield tended to decline under poplar as well as in the open area; the reduction was, however, less in the latter treatment. Wheat yield was generally higher when leaf mulch was not used (Anonymous, 1990).

Chrysopogon montanus, Heteropogon contortus, Themeda anathera, Dichanthium annulatum, Capillipedium huegelii and *Eragrostis* spp. are some of the dominant plant components in natural grassland ecosystems. *Setaria*, white clover and alfalfa are promising species for increasing the productivity of natural grasslands. Agri-silviculture, silvi-pastoral and horti-silviculture agroforestry systems require immediate attention in this zone.

(c) High-hill dry temperate zone

Annual precipitation is below 20 cm, most of which is received as snow during winter. Cultivation is done during summer and autumn with the help of irrigation. The region is ideally suited for the production of quality seed of potato, chicory and temperate vegetables. Wheat, buckwheat, barley and amaranth are cultivated for domestic use. The main stress of the farming community is on the growing of fruit and rearing of sheep and goat. Almond, prunes, drying variety of apricot, grape, chilgoza, apple, hops, pistachionut, sarda melon are important fruit crops of the area. The natural vegetation consists of *Juniperus macropoda, Betula, Populus, Salix, Hypophoea rhamnoides, Ribes, Ephedra, Artemisia, Astragalas, Cymbopogon*, etc. Trees of *Robinia, Ailanthus altissima, Juglans, Populus, Salix, Melia* etc. are often planted for fodder and fuel purposes (Verma *et al.*, 1989).

Tewari (1986) analysed the economics of various agroforestry system after introducing *Leucaena leucocephala* in the existing cropping system of lower hills of Himachal Pradesh. The study showed that per hectare net return can be increased to three times from Rs. 208.30 to Rs. 598.62 per year by adopting agri-silviculture instead of the existing cropping systems. However, the adoption of agro-silvi-pastoral system increased the net returns to Rs. 1350.60 per hectare. The integration of trees with crops/pastures or vice versa is thus more economical than cultivation of agricultural or horticultural crops alone.

Mazumdar (1991) studied different agroforestry systems around the university at Nauni (Solan). The gross annual return from agriculture (including vegetables crops) alone was Rs. 88981 per hectare which increased to Rs. 91577 per hectare in horti-agriculture system. In contrast horti-silvi-pastoral and grasslands had annual returns of Rs. 13870 and Rs. 2910 per hectare only.

A good deal of work embracing different facets of various agroforestry systems is underway in Dr. Y.S. Parmar University of Horticulture and

Forestry, Nauni (Solan).

B. Eastern Himalayas

A classical example of agroforestry in the eastern Himalayas is the cultivation of large cardamom (*Amomum subulatum*) under tree shade at 600 to 2000 m altitude. Large cardamom is a shade loving plant. *Alnus nepalensis* is the predominant tree in its plantations although *Schima wallichii, Maesa chesia* etc. and a host of fodder species, viz. *Ficus* spp., *Machilus edulis, Celtis chinnamomea, Eurya japonica, Saurauvia nepalensis, Pieris ovalifolia, Dendrocalamus sikkimensis, Arundinaria hookeriana* etc. are also common in the plantations. *Alnus nepalensis* is a nitrogen-fixing deciduous diazatroph. The leaf litter decomposes easily and helps in nourishing the crop on sloppy lands where fertilizer application is a difficult proposition and its efficiency of use is very low due to heavy rainfall. There is a cyclic order of felling older trees and allowing younger ones to grow. If older trees having more than 16 cm basal diameter are not cut, their horizontally spreading roots get sufficiently thickened and lift the clumps of large cardamom (Singh *et. al.*, 1989). The profitability of the system can be increased further by rearing honeybees in the plantation for better pollination of large cardamom and greater production of honey (John and Mathew, 1979 and Jacob, 1980).

Fodder and fuel trees are found growing on the terrace risers of upland crop fields, along farm boundaries and water springs passing through the farmlands. Crops grown on the upland terraces are maize, wheat, barley, buckwheat, vegetables, black gram and ricebean (Singh, 1987). Fodder trees are systematically planted on terrace risers at 4-5 metre intervals. Some of the fodder trees are retained from natural regeneration when the land is terraced for cultivation. Most of the upland crop fields look like a mosaic due to various trees on them. *Ficus* spp., *Artocarpus lakoocha* and *Litsea polyantha* are the most popular fodder trees on farmlands. Fodder tree leaves contain good protein and calcium levels for milch cows and are important forage materials available throughout the year. Few more trees viz. *Grewia optiva, Robinia pseudacacia* and *Leucaena leucocephala* have now also been included for evaluation under the All-India Coordinated Research Project on Agroforestry at Sikkim Centre Tadong, Gangtok (Singh *et al.*, 1989).

Sechium edule, a vegetable crop, is raised near *Schima wallichii* which offers support and aerial space for the vines of the former to grow and spread over the entire canopy. This puts luxuriant vegetative growth with abundant fruits and is often fed to pigs (Venugopal, 1986).

Mandarin (*Citrus reticulata*) orchards are intensively intercropped with many species, cereals, pulses and vegetable (Singh, 1987). Ginger is the most remunerative crop. The large doses of farm yard manure given to ginger benefit the mandarin tree as well. The low productivity of intercrops is in consonance with the subsistence farming practised in this region.

In the high hills of temperate zone, apple is intercropped with potato, barley, radish, cabbage and turnip (Singh *et al.*, 1989). At high elevations of Meghalaya cash crops such as potato, rice, maize and ginger are grown in monoculture or mixed culture with *Pinus kesiya*. Lopped branches of the trees species are placed on the soil ridges and burnt partly to form an excellent bed for growing of agricultural crops. Two or three relay crops are taken over one year (Mishra and Ramakrishnan, 1981).

Bamboos cover extensive areas in the eastern Himalayas. Owing to their multifarious utility and excellent soil binding property they are preserved in agricultural fields either on borders or in strips. Apart from the human food, bamboo leaves serve as an excellent winter fodder for goats (Singh *et al.*, 1981). Agricultural lands shaded by *Bambusa nutans* can be effectively utilised by growing ginger, turmeric, large cardamom, orchard grass and dinanath grass. Grasses play an important role in reducing soil erosion, run-off and nutrient losses. Amliso (*Thysanolaena agrostis* Syn. *T. maxima*) has been identified as the most promising shade tolerant grass having multiuse values (leaves for fodder, stem for pulp and spike for broom) and is best suited for silvi-pastoral systems. For uncultivated sloppy wastelands in mid hills, silvi-pastoral systems based on nevaro (*Ficus hookerii*) and existing forest (*Alnus nepalensis* and *Schima wallichii* mixed forest) have been developed. These systems have a carrying capacity of 9-15 local goats per hectare. Secondary productivity of nevaro-nandi setaria system showed weight gain of 69 + 12.41 g per day under growth-cum-metabolism trial on local goats (Singh *et al.*, 1989).

Roy and Verma (1976) suggested the cultivation of wheat grass (*Agropyron semicostatum*), colonial bent (*Agrostis tenuis*), doob grass (*Cynodon dactylon*), orchard grass (*Dactylis glomerata*) and weeping love-grass (*Eragrostis curvula*) on the terraces left after cropping. With the introduction of nutritive grasses and legumes, animal husbandry can form a source of subsidiary income for tribal people.

Several degraded pastures and grasslands exist in the temperate zone of eastern Himalayas. It is important that suitable silvi-pastoral combinations may be developed by planting suitable multipurpose trees (Khosla and Toky, 1991).

Borthakur *et al.* (1981) proposed a model for land use in the hilly terrain as an alternative to shifting agriculture. Under this model, the upper one-third portion of the hillock is left as intact forest, middle one-third is used for growing fodder and fruit species and the bottom one-third is terraced and made into permanent agricultural fields. Owing to the raising of horticultural crops, the farmers will have along term interest in the land and round-the-year cropping with good yield may be obtained from the terraces. However, the model needs to be tested by the farmers over considerable period of time.

Future Prospects

Agriculture and forestry are inherently more complex systems than the typical industrial plant. The integration of the two in agroforestry further increases the complexity. An important characteristic of an ideal agroforestry system is the existence of complementary (preferably) or supplementary interaction among the various components. But, instead, intercomponent competition is the principal interaction in the temperate region of India, particularly in the western Himalayas. This is attributable to the adoption of defective crop-tree combinations. In order to explore the optimum natural resources of radiant energy, water, nutrients and space, plant species of diverse growth habits, rooting patterns and requirements should be put together mimicking the natural multistorey ecosystems. Intensive research effort is, therefore, needed to identify the appropriate combinations of components. For instance in the Sikkim Himalaya *Alnus nepalensis* and *Schima wallichii* could be widely spaced (20 m to 30 m apart), and a second tier of fast growing high value fodder trees as *Ficus* spp. should be planted at shorter (10 m to 15 m) intervals in between them. High yielding temperate perennial grasses of good fodder value as grassland manava, perennial rye grass, thin napier, Guatemala grass, weeping love-grass, orchard grass could occupy the ground cover. More shade tolerant crops and varieties need to be identified and evolved for growing between the rows of closely spaced fodder trees. Several medicinal plant species thrive in shady environments in the forest and can, therefore, be taken up for cultivation in tree shade.

Exotic fuel-cum-fodder trees such as *Robinia pseudacacia* and promising hybrids of *Populus* spp. should be tried and evaluated for their performance (Khosla and Chauhan, 1983). Some lesser - known fast growing fodder species like *Ulmus laevigata* need to be tried in different areas for their adoption on a larger scale. Tropical species e.g. *Sesbania, Leucaena, Glircidia* etc. may prove successful in some pockets in the Himalayas. Nitrogen-fixing trees and forage crops need to be used on a larger scale for nitrogen enrichment in the soil.

Owing to the severe scarcity of fodder and wood, the farmers lop the trees very heavily, leaving behind a deformed stem with very few buds and branches. This plays havoc with the vigour and yield of the tree. The lopping and pollarding techniques need to be standardised.

An efficient network of extension workers is also required for disseminating the information from scientists to the rural folk and for providing feedback to the scientists.

The success of agroforestry has also been obstructed by the shortage of inputs and financial resources. Introduction of agroforestry increases the credit requirement. Tewari *et al.* (1990) recommended that more credit facilities should be extended to the small farmers who wish to adopt a more productive but expensive technology.

The weak scientific knowledge about socio-economic and technical aspects of agroforestry, the complexity of agroforestry system and the value of local farmers' experience make on-farm research unusually important in agroforestry. In-depth diagnostic and descriptive research is needed prior to the design of long-term experimental programmes. A strong parallel programme of technology testing with farmers is needed to complement the selective strategic research on component response and interactions in researcher-controlled plots. This approach requires active institutional collaboration, strengthening of research capacity for technology testing in extension programmes and expanded training for researchers in on-farm research methods (Scherr, 1991).

Summary

Agroforestry is an age-old practice in the temperate Himalayas. Fodder and fuel species are planted in and around the farms under the agri-silvicultural, horti-silvicultural, agri-horticultural, silvi-pastoral and agro-silvi-pastoral systems. Trees are not properly protected, replanted or managed. Agroforestry practices are generally rudimentary and primitive, economic return from these are very low. There is a great potential for the improvement of the traditionally managed systems for realizing the real production potential of agroforestry systems.

REFERENCES

Anonymous. 1987. Annual Research Report, 1986-87, Dept. of Silviculture and Agroforestry, UHF, Solan.

Anonymous. 1988. Annual Research Report, 1987-88, Dept. of Silviculture and Agroforestry, UHF, Solan.

Anonymous. 1989a. Research Report. 1987-89, AICRP on Agroforestry, Dept. of Silviculture and Agroforestry, UHF, Solan.

Anonymous. 1989b. Annual Research Report, 1988-89, Dept. of Silviculture and Agroforestry, UHF, Solan.

Anonymous. 1990. Annual Research Report, 1989-90, Dept. of Silviculture and Agroforestry, UHF, Solan.

Anonymous. 1991. Annual Research Report. 1990-91, Dept. of Silviculture and Agroforestry, UHF, Solan.

Borthakur, B.N., Prasad, R.N., Ghosh, S.P., Singh, A., Awasthi, R.P., Rai, R.N., Verma, A., Datta, H.M., Sachan, J.N. and Singh, M.D. 1981. Agroforestry - based farming system as an alternative to jhuming. In: Proc. 'Agroforestry Seminar', ICAR, New Delhi. pp. 109-31.

Dadhwal, K.S. Narain, P. and Dhyani, S.K. 1989. Agroforestry systems in the Garhwal Himalaya of India. Agroforestry Systems. 7: 213-225.

Gupta, R.K. 1983. Watershed management and agroforestry in western Himalaya. Lecture delivered during Summer Institute on Agroforestry, HPKV, Solan. pp. 1-40.

Gupta, R.K., Arora, V.K. and Shukla, D.L. 1982. Know your peach cultivar for Doon Valley. Indian Hort. 2-4.

Jacob, K.J. 1980. Pollination — most decisive factor in cardamom production. Cardamom 12(8): 3-9.

John, J.M. 1979. Large cardamom in India. Cardamom 11(10): 13-20.

Khosla, P.K. and Chauhan, P.S. 1983. Agroforestry resources in western Himalaya. In: Summer Institute on Agroforestry in western Himalaya. June 13 - July 7, HPKVV, Solan, 78+iv p.

Khosla, P.K. and Toky, O.P. 1991. Agroforestry systems in the Himalaya. Personal communication.

Mazumdar, H.K. 1991. Biomass productivity and nutrient budgeting in different agroecosystems. Ph.D. thesis, UHF, Solan. pp. 16-107.

Mishra, B.K. and Ramakrishnan, P.S. 1981. The economic yield and energy efficiency of hill agro-ecosystem at higher elevations of Meghalaya in north-eastern India. Acta Oecologia/Oecologia Applicata. 2: 369-89.

Mudgal, V.D. and Pradhan, K. 1988. Animal feed resources and current patterns of utilisation in India. In: Non-conventional feed resources and fibrous agricultural residues (Ed. Devendra, C.) IDRC/ICAR. pp. 139-146.

Puri, D.N. 1990. Production potential of degraded lands with multipurpose tree species. In: Multipurpose tree species research for small farms : strategies and methods. pp. 184-87.

Roy, D.J. and Verma, A. 1976. Animal husbandry as a subsidiary source of economy for jhumians. In: Shifting cultivation in north-east India. North-East India Council for Social Science Research, New Delhi. pp. 7-51.

Scherr, S.J. 1991. On - farm research : The challenge of agroforestry. Agroforestry Systems 15: 95-110.

Shamet, G.S. and Puri, S. 1986. Rehabilitation of denuded lands through watershed management and agroforestry approaches with special reference to Shivalik hills. In: Agroforestry Systems : A new challenge (Eds. Khosla, P.K., Puri, S. and Khurana, D.K.). Indian Society of Tree Scientists, UHF, Solan. pp. 81-91.

Singh, G.B., Venugopal, K., Balaraman, N. and Beniwal, R.K. 1981. Agriculture resource utilization in Sikkim - Present status and future prospects. In: Meeting on overall land and water use planning for Sikkim. April 8-17, 1981.

Singh, K.A. 1987. Annual Report, 1987. AICRP on Agroforestry. ICAR Research Complex for NEH Region, Sikkim Centre, Gangtok. pp. 12-38.

Singh, K.A, Pradhan, I.P., Rai, R.N. and Awasthi, R.P. 1989. Agroforestry research status in eastern Himalayas. In: Agroforestry systems in India - Research and Development. Indian Society of Agronomy, IARI, New Delhi. pp. 47-63.

Singh, R.V. 1986. People's participation in social forestry programmes in Himachal Pradesh. Occasional paper 10, FRI and Colleges, Dehra Dun.

Swarup, R. and Tewari, S.C. 1988. Farmers' dependence on forests for fodder, fuel and timber in hills of Himachal Pradesh. In: Trends in tree sciences (Eds. Khosla, P.K. and Sehgal, R.N.). Indian Society of Tree Scientists, UHF, Nauni. pp. 6-11.

Tewari, S.C. 1986. Ameliorating farm income by agroforestry on marginal and submarginal lands in lower hills of Himachal Pradesh. In: Agroforestry systems - A new challenge (Eds. Khosla, P.K., Puri, S. and Khurana, D.K.). Indian Society of Tree Scientists, UHF, Solan. pp. 185-90.

Tewari, S.C., Guleria, J. and Sharma, R.K. 1990. Capital and credit needs of western Himalaya farmers for agroforestry. Agri. Situation in India. 45(9): 589-91.

Venugopal, K. 1986. Prospects of agroforestry in Sikkim. In: Agroforestry systems - A new challenge (Eds. Khosla, P.K., Puri, S. and Khurana, D.K.). Indian Society of Tree Scientists, UHF, Solan. pp. 69-74.

Verma, K.S., Mishra, V.K. and Sharma, S.K. 1989. Agroforestry in Himachal Pradesh - a case study. In: Agroforestry systems in India - Research and development (Eds. Singh, R.P., Ahlawat, I.P.S. and Saran, G.). Indian Society of Agronomy, IARI, New Delhi. pp. 64-70.

Chapter 10

Agroforestry and Watershed Management

P.K. Sen Sarma* and R. Jha**

Introduction

Forests in India cover an area of 75.18 million ha as per official records. However, due to various non-forestry uses such as usurping for agricultural practices, dams, water reservoir for hydro-electricity, building of roads and other communication network etc., the forest area has been reduced to 64.20 million ha according to the satellite imagery and in the major area forest canopy cover is less than 40 per cent. The degradation of forests coupled with increasing population of human kind and cattle, good and services are progressively becoming scarce leaving aside amelioration of ecological and environmental hazards. There is an acute shortage of fuelwood and raw material for wood based industries. The fuelwood shortage has already assumed an alarming proportion. As against the projected demand of 240 million tones, the current supply is only 40 million tones. Keeping these in view the National Forest Policy 1988 enunciated *inter alia* the following :

- Maintenance of environmental stability through preservation and where necessary, restoration of the ecological balance that has been adversely disturbed by serious depletion of the forests of the country.
- Increasing substantially the forest/Tree cover in the country through massive afforestation and social forestry programmes especially on all denuded, degraded and unproductive lands.
- Meeting the requirements of fuelwood, fodder, minor forest produce and small timber for the rural and tribal populations.

* *S.F.S.O., Shillong.*
** *R.A.U., Pusa, (Bihar).*

- A massive need-based and time bound programme of afforestation and tree planting with particular emphasis on fuelwood and fodder development on all degraded and denuded lands in the country, whether forests or non-forest land, is a national imperative.
- Village and community lands, including those on foreshores and environs of tanks, not required for other productive use, should be taken up for the development of tree crops and fodder resources.

Landuse pattern in India

Landuse patterns as given in Statistical Handbooks issued by various States show the following :

Sl. No.	*Landuse*	*Area in million ha*	*Percentage of total area*
1.	Agriculture (cultivable land)	154.70	47.0
2.	Forests (Area officially recorded as forests with or without tree cover)	75.18	22.8
3.	Permanent pastures and other grazing lands	12.15	3.7
4.	Land under cultivable tree crops and groves	3.91	1.3
5.	Cultivable wasteland	16.64	5.1
6.	Land under other non-agricultural uses.	17.53	5.3
7.	Barren and wastelands	24.60	7.5
8.	Area for which no returns exists	24.09	7.3
		328.80	100%

Agroforestry

Agroforestry is defined as sustainable land management system which increases the yield of land, combines the production of crops (including tree crops) or animals simultaneously or sequentially, on the same unit of land and applies the management practices of local farmers. This is thus, a generic term which encompasses a variety of land use patterns incorporating trees as one of the components.

The concept thus eschews the artificial dichotomy of agriculture and forestry includes agri-silviculture, silvi-pastoral, agri-silvi-pastoral, horti-silvi-pastoral, silvi-agriculture, silvi-lac-cultivation and silvi-tassar culture etc. According to F.A.O., the first four systems are the immediate requirements for the successful agroforestry programmes. This can be done in two broad manners (i) integrating crop cultivation with tree growing, (ii) and/or growing trees with crop production.

In the agri-silviculture system, agriculture crops including legumes are grown alongwith tree crops in the agri-cultural land following farmers practices. But growing forest trees alongwith agricultural crops in forests for a temporary period is an old system known as "Taungya system" and the two are not synonymous.

In the silvi-pastoral and agri-silvi-pastoral systems, land is managed in such a manner that trees, crops, grasses and seasonal forage crops are grown simultaneously. In the Horti-silvi-pastoral system, seasonal horticultural crops (mainly vegetables) alongwith forest tree crops and grasses and legumes are grown.

Watershed, Shifting Cultivation and Agroforestry

Several tribes in India practices shifting cultivation or jhum in watershed areas. In this practice, areas in forests selected for *jhuming* is clear felled, timber and fuelwood are taken out and sold, and the area is put under fire in order to enhance the fertility of the soil. In earlier years, the fallow periods between the two jhums was long enough to permit natural regeneration of original vegetation and restoration of soil fertility. But on account of population pressure, the fallow period is now too short to permit natural regeneration of vegetation with resultant soil infertility. These areas are also prone to heavy soil erosion. It has been calculated that in Meghalaya with 50-60 per cent slopes, soil loss is 147 tonnes/ha in first year *jhum*, 170 tonnes/ha in the second year, *jhum* and only 30 tonnes/ha in abandoned *jhum*. This erosion rate is much higher than the acceptable limit. The rate of soil erosion under different cropping patterns is as follows :-

Forests 24.5m^3/Km^2/Yr, tea plantation 483 m^3/Km^2/Yr and vegetable garden 732 m^3/Km^2/Yr. This clearly shows that shallow rooted crops cause higher soil losses than in case of deep rooted tree crops.

Surface runoff in shifting cultivation area is about 111.4 mm. Partial terracing and complete terracing reduce it to 81.4 mm, and 32.8 mm. respectively. A strip of trees planted along contours serves the same purpose as terracing. The shifting cultivators can be weaned away from the harmful our practice of shifting cultivation by resorting to agroforestry practices which are in conformity to their cultural milieau.

Suitability characteristics of tree species for agroforestry

Tree species intended to be grown in conjunction with crops should *inter alia* have the following characteristics :

(a) amenability to early espacement.

(b) Self-pruning property or capacity to withstand heavy artificial pruning.

(c) Low crown-bole diameter ratio.

(d) Light branching.

(e) Tolerance to side-shading.

(f) Appropriate phyllotaxi for penetration of light to ground.

(g) Litter fall and litter decomposition rate having positive effect on soil.

(h) Absence of competition at root-zone level.

(i) Efficient nutrition pumping.

(j) Amenability to silvicultural practices.

To summarise, architecture, morphology, phenology, root distribution, root spreading and activity of woody perennials must be taken into consideration while selecting tree species for incorporation in agricultural system.

Watershed Management

Availability of water in a given soil environment is a critical factor which is adversely affected through surface run-off, erosion, and siltation, due to loss of beneficial plant cover, water management is best done on a natural unit, watershed which a distinct drainage basin of any small or big water course or stream, nullah etc. The rains falling over watershed areas will flow through only one point of exit of the whole watershed. In other words, the area will be drained by only by one stream or water source. Thus, this broadly encompass availability of entire water resource which can be utilised through dug wells, tube wells, small farm ponds, bigger tanks or reservoirs, etc. Siltation prevention is one of the critical factors for the proper management of the water reservoir. About 914 priority watershed of the flood prone areas and 4246 micro-watersheds had been identified in 1982-83. Stabilisation of watersheds, is recommended to be achieved through massive afforestation programmes. Agroforestry systems mimic the major functions of forestry in respect of watershed afforestation. Thus adoption of agroforestry practices is expected to provide the same benefits that are achieved through afforestation.

It is well accepted that aim of management of watershed should be higher productivity of the land in the catchment areas in addition to its traditional aim to prevent surface run-off and soil erosion. The aim of agroforestry, is to optimise positive interactions between various biological components like trees, shrubs crops and animals, and between these components and the physical environment so as to obtain a higher, more diversified and sustainable production system from the land than is possible with other form of land use under the prevailing ecological and socio-economic conditions. Agroforestry approach to watershed management brings about the potential role of trees/shrubs to alleviate some of the physical and economic constraints. Agroforestry mimics the beneficial aspects of forests, horticulture and agriculture in reducing soil erosion, surface run-off, siltation loads of water courses, and nutrient loss of the soil. Further, agroforestry system brings about less solar radiation, low soil temperature, higher ambient soil moisture in summer months, improved soil organic matter, soil pH, lower incidence of pests and diseases of crops, and overall improvement in macro and micro-climate. All these are conducive to the holistic management of watershed through judicious admixture of various biological components that would promote an integrated management system so that the system produces goods and services to fulfil the demand of the society.

REFERENCES

Prasad, R.B. (1984). Technology for dryland agriculture on watershed basis. *Proceeding* Dryland farming and watershed management. (Mohsin, M.A. and Pd. R.D., Eds.) B.A.U. Ranchi, pp. 72-83.

Mohsin, M.A. (1984). Forestry and watershed management as interlocked system. *Proceeding* social forestry and watershed management, (Muhammad, S. Eds.), B.A.U. Ranchi, pp. 110-115.

Mohsin, M.A. (1989). Watershed and its management, K.V.K. Souvenir, Ramkrishna Mission, Ranchi, pp. 33-39.

Srivastra, V.C. (1984). Water management its importance in dryland agriculture. *Proceeding*, Dryland farming and watershed management, (Mohsin, M.A. & Prasad, R.B., Eds.). B.A.U. Ranchi pp. 162-176.

Sinha, A.K. (1984). Approved practices of soil conservation for watershed management. *Proceeding* Dryland farming and watershed management, (Mohsin, M.A. and Pd. R.B., Eds.) B.A.U. Ranchi.

under bare soil situations. The co-efficient of runoff, C, in the modified Rational formula, also has been observed to be lower by 20 to 60 per cent on forest than on cultivated land, and from 11 to 30 per cent lower on forest than on pasture land under Dehra Dun and Chandigarh conditions. At Chandigarh, observation on sediment yield into a farm pond showed that soil conservation measures in its catchment reduced the sediment yield from 0.71 to 0.07 ha. m/sq.km over seven years. At Dehra Dun, sediment yields from a 14 ha. watershed with cultivated terraces and unterraced land was 62 tonnes/ha over a 13 years period, as compared to 15 tonnes/ha from a 2.2 ha forested catchment. Similar results are available for Hazaribagh soil and climatic conditions which are based on researches conducted at the Research Station of the Damodar Valley Corporation and at the Soil conservation Research Station of the Department of Agriculture, Bihar.

In spite of all such revealing research results, the fact remains that increased population pressure has resulted in an upward extension of agricultural land at the cost of is being very severely degraded through lopping for fuel and fodder, and through overgrazing. The estimated requirement of food and fuel by the turn of century is expected to be as high as 225 m.t. and 250 m.t., respectively.

Factors Leading to Land Degradation

The problem of land degradation is evidently based on inter-relationship between several factors such as :

(a) traditional attitude to cattle;

(b) need for organic fertilizer or fuel in the form of dung;

(c) fragmentation of farm holdings;

(d) perceived need of farmers for their own draft animals;

(e) traditional attitude of farmers to, and right in, communal and government owned forest land.

Specific Problems in Forestry

The specific problems in the field of forestry have been identified as :

(a) An increasing need for timber, fuel and fodder, resulting in degradation of forest land by excessive tree lopping and overgrazing.

(b) The upward extension of terraces for cultivation, on steep lands which should be under permanent tree cover (forest or orchard).

(c) The failure of forest departments to take into account the needs of the local population in drawing up forest management plans (e.g. location and extent of closure, species for reforestation).

(d) Inadequate contact between Forest Department and villagers through extension, both for technical advice on communal and private lands, and for changing attitudes towards communal and government owned lands.

(e) Peacemeal treatment of denuded catchments by a number of agencies working independently, and lack of coordination between territorial and soil conservation branches of the state forest department.

(f) Inadequate drainage and soil conservation measures in the construction and maintenance of forest roads.

(g) The frequent use of fire to encourage new grass growth on fresh lands, resulting in damage to seedlings, removal of forest litter and in the long term loss of soil nutrients.

(h) Emphasis placed on the commercial role of forests for timber production on sites which may be better used for the production of fodder and fuel, or for protection on severely erodible sites.

(i) Inadequate factual information on which to base management decisions.

Forestry related activities in watersheds

The objectives of forestry-related activities in the watersheds should be :

(a) The establishment and maintenance of forest, brush or grass cover for soil conservation and water retention.

(b) The provision of fuel wood, fodder and timber trees (for agricultural implements) of desirable species in sufficient quantity and at a convenient distance from villages where it will be utilised.

(c) The establishment, management, utilisation and regeneration of commercial forests on appropriate lands which are not required for protective, soil conservation or fuelwood/fodder purposes. Trees for people would have to be overriding criteria.

(d) Effective regeneration of logged and degraded lands.

(e) Establishment of nurseries.

(f) Stabilisation of roadworks.

(g) Provision of fire protection measures, including detection, suppression and fuel reduction (control burning).

(h) Provision of an effective extension service to improve cooperation with villagers.

(i) Initiation of an effective research programme.

(j) Provision of effective training to forestry personnel, particularly at forest-guard and ranger's level.

(k) Improvement of local employment opportunities.

A few words on some specific priority components

(a) Afforestation of denuded lands

It can be very well visualised that the most severely denuded areas are in the civil and private lands rather than in reserved forests. These are the lands which are grouped under class V to Class VII lands in the land capability classification in a Watershed Management Plan. Reforestation, wherever recommended, should be with species most suitable for fodder and fuel production. Trees species, which have fast growing habit must be chosen which will provide for villagers' needs. Some of the fodder species which have been recommended for silvipasture system include *Bauhinia variegata* (kachnar), *Toona citiata* (tun), *Sapindus mukoorossi* (Kusum) *Dichrostachys mutens* (ban), *Robinia pseudocacia, Albizzia amara, Acacia tortilis, Morus alba, Grewia oppositifolia, Gmelina arborea* and *Leucaena leucocephala.*

Before planting can be undertaken, closure, fencing and subsequent control of access is an obvious prerequisite. Intensive extension work with villagers must be undertaken in order to ensure their cooperation, and incentives such as provision of fodder and fuel from other adjacent lands must be provided. Normally phased closure of 10 to 20 ha. units is recommended, as this is supposed to be an manageable size for fencing and planting, and does not remove too much land from grazing at one time. Fast growing fuelwood species (e.g. Eucalyptus) could be planted at close spacing (2m), four or five rows deep, around the periphery of closed areas, so as to form a 'living fence' after some years. Controlled grass cutting should be encouraged in closed areas one year after planting, and controlled grazing can generally be permitted after 5 to 10 years depending on species.

The prospects of tree farming on private and communal lands should be seriously considered. In places like Himachal Pradesh, and in many other parts of the country, farmers have been persuaded to convert much of their agricultural land of low productivity to the cultivation of Khair (*Acacia catechu*) whose wood is selling at Rs. 100 or even more per quintal, at some places. At many places fast growing Eucalyptus has been successful for providing fuelwood or for sale as 'cash crop' for pulp, on rotation as short as 5 or 6 years, on productive soils. At places agroforestry, social forestry or even agro-silvipastoral system could be adopted. Agroforestry is an inter-disciplinary approach to systems of land use based on forestry, agriculture, animal husbandry including pastures, acquaculture and fisheries. An association of trees with annual or perennial crops is generally identified as a system belonging to agrisilviculture technique (or silviagricultural technique). When animals, domesticated or wild, are also associated within the system it is termed as silvipastoral technique. Where all the three elements are simultaneously associated in the pattern of land use, these have been called 'agrosilvipastoral system'.

(b) Establishment of Nurseries

Seedlings are required for any planting which is carried out, whether for reforestation, afforestation, gully control or other soil stabilisation work. At many placed the nurseries are located at permanent sites and seedlings are transported over relatively long distances by truck. Where transportation costs are heavy, this may not be desirable. In such cases temporary nurseries are preferred as close as possible to the planting sites. In many places, the average nursery size ranges between 1 and 2 hectare and seedling production is about 100,000 per hectare. A one hectare nursery thus, may produce sufficient planting stock for about 100 ha. @ 1000 trees per hectare, or 250 ha. @ 400 trees per hectare. A 12 ha planting unit would therefore only require a nursery of about 1000 sq. meter. However, if a number of planting units are established within 2 to 3 km of each other, a single nursery to supply several units may be preferable. Temporary nurseries tend to be less than 1 ha. in size.

(c) Firewood supply and distribution

The need and feasibility of establishing specific firewood plantations will vary from one catchment to another. Where there is serious shortage of fuel in the surrounding villages, and there is no suitable alternate sources in existing forest land, closure and planting with fuelwood species should certainly be considered. In general, however, multipurpose plantations which supply fuel, grass and tree fodder would be preferable. Some fast growing species of fuelwood plantation like, *Robinia, Salix, Populus, Eucalyptus,* and *Acacia* may be considered. There should be provision of firewood depots also.

Where villagers have been fulfilling their fuelwood needs free of charge from surrounding forests, they are not likely to respond favourable to having to purchase the fuel from Government Depots. *However, if closure of degraded forest is to be successful free or heavily subsidised provision of an alternative source of fuelwood might form a component of the watershed project.*

(d) Villagers participation and extension work

The purpose of extension work in the Watershed with special reference to forestry would be to :

(i) Take villagers' needs into account when drawing up management plans for lands under the control of the Forest Department.

(ii) Improve the management of commercial and private forests by the provision of technical assistance.

(iii) Change villagers' attitudes to the protection and use of government land.

Concluding remarks

In the watersheds and catchments, the maintenance of forests, or the development of the same in degraded forest areas, can be taken up either as a programme of commercial forestry, as suggested in the report of the National Commission on Agriculture, entitled "Interim Report on Production Forestry-Man-made forests" or as a Social Forestry Programme for improving fuelwood and timber supply to the villages as suggested in another "Interim Report on Social forestry". However, one of the main difficulties in the proper conservation of forest lands in catchment areas is the presence of extensive rights and privileges of villagers for grazing and for collection and use of forest produce. The National Commission on Agriculture recommended that "it would be necessary not only to regulate their rights and privileges but also to supply their requirements of forest produce in such a manner that the important role of forests in conserving soil and moisture is not lost sight of". In the watershed, free or heavily subsidised provision of an alternate source of fuelwood shall have to be considered.

The development of agro-forestry, or even agri-silvi-pastoral system is very much essential in any programme of watershed management. In other words, proper Watershed Management is highly dependent on forestry practices and the two may be considered as an inseperable, interwoven and completely interlocked system.

Chapter 12

Approved Practices of Soil Conservation for Watershed Management

*A.K. Sinha**

Importance of soil conservation

The soil is that layer of the earth's surface which is composed of mineral and organic matter and is capable of sustaining plant life. Quantity, quality and frequency of economic output from any piece of land are conditioned in a large measure by the nature and characteristics of the soil in the upper layer, suitable depth of which takes about hundred years or more for its natural formation. The loss of this valuable top soil results in a deterioration of the quality of land so much so that its use pattern has to undergo a complete change for the poorer. The loss of top soil is a slow, gradual process which can, however, be accelerated by injudicious or ignorant human action in the form of neglectful or faulty methods of husbandry that tend to interfere with the cyclical process of nature and in particular, accentuate the destructive part of it. During the period that the top is being lost, the affected land suffers from regular process of depletion of productivity. This can however be replenished to some extent by adopting of appropriate soil conservation and antierosion measures.

Soil conservation in its broad sense thus means not only adoption of measures for protecting lands from the process of erosion and deterioration in quality but include also such measures which are taken to upgrade the affected land. It does not, however, mean complete stoppage of all exploitation and use of lands, but the regulation of use in a manner which is most economical from the point of view of the individual and the society over time. Soil Conservation constitutes a part an important part of the wider issue of conservation of natural resources.

**Deptt. of Agril. Engineering, Faculty of Agriculture, Birsa Agril, Univ. Ranchi.*

The historical and archaeological evidences show that the land resources are exhaustible and nations that have not taken care of their lands have had to face extinction. Growing incidence of flood and famine in recent years are the result of misuse of land and water resources. Quantitative data to show the magnitude cf the deterioration of land and depletion of productivity that have taken place in the past are not much available. There are, however, some broad estimates available which shows the granity of this menance in India.

A survey in an area in the Chotanagpur region of Bihar which was subject to active arosion showed that during about 45 years since the beginning of this century, approximately 17 per cent of the surveyed area have been lost to cultivation through gully erosion. This gives a measure of the rate of deterioration in the quality of land.

In an another estimate it was observed that in the Sholapur district of Maharashtra State, 17 per cent of the land having medium soil depth between 18 to 36 inches deteriorated into shallower soils (less than 18 inches) in the years from 1870 to 1945. It was also observed at the Dry Farming Station at Sholapur that in medium soils average grain yield of Jowar was 169 Lbs per acre while in shallow soils it was only 69 Lbs per acre.

Under the auspicious of the Central Board of Irrigation and Power a systematic study of the rate of siltation in selected reservoirs i.e. Bhakra, Panchat Hill, Maitlan, Mayurakshi and Tungbhadra, revealed that due to heavy erosion in the catchment area, the actual rate of siltation was 2 to 3 times more than the assumed designed load. The resultant effect is the lessering of the life of the reservoir which will cause great financial loss to the nation.

Dr. R.L. Singh, Director, National Park, Dudhwa, has recently estimated that India is loosing six crores tonnes of soil annually as a result of erosion and the minerals washed away with it will cost Rs. 700 crores on the present day market value.

These examples amply projects the gravity of the menance and calls for urgent remedial measures.

Objective of soil conservation

The objective of soil conservation is the selection on the basis of requirements and adoption of long lasting as well as recurring measures with a view to prevent the ultimate deterioration of land and providing for its exploitation and use in a manner designed to reduce the loss of soil and conserve productivity of land to the point most economical in the interest of the society as well as the individual. It aims to regulate the present land use in a manner which may result in a progressive rise in productivity of land and also preserve and enliance its quality for the posterity. The method of achieving this objective, in simple but scientific words, is to use each piece of land according to its capability, and treat it according to its need. It may require improvements in the prevailling practices and management of land in a number of directions and over a period of time on complete change over to a new management programme.

Common soil conservation practices

It is of utmost importance that Soil Conservation Programme should be carried out on the watershed basis i.e. the unit of operation should essentially be a watershed. The land under a watershed may be under different types of uses like forest, pasture, cultivation, habitation etc., with varieties of combinations of soil, moisture status topography, vegetation etc. Thus each type of land may require specific treatment. It is also important that for the success of such programme the whole area, irrespective of the land use, should be treated simultaneously.

Therefore, soil conservation programme in its true sense is a package programme, all items of which, to the extent they are necessary are together essential for obtaining sustained increase in the productivity of land and achieve the long term effect. Since a number of items are involved, the benefits of each or of the whole are difficult to demonstrate, which certainly is a disadvantage from the point of view of extension education.

The soil conservation measures fall under three broad heads:-

(i) *Agronomic* - Cover crop, contour cultivation, strip cropping, crop rotation, tillage, mulching, manuring etc.

(ii) *Forestry and Pasture management*

(iii) *Engineering* - Terracing, bunding, grass, waterways gully plugging, spillways, silt detension dame etc.

But it must be clearly understand that these measures are not independent rather they are supplemental to each other. For best results they should form integral part of the soil conservation project to strengthen the effects of each other as per need.

The important measures suitable for this are being discussed in this paper :-

Contour cultivation

One of the most successful method of reducing runoff and erosion is contour farming i.e. conducting all farming operations on contour or across the slope. Light showers are absorbed readily in the plough furrows and help moisture, conservations, whereas in the event of heavy rainfall contour furrows reduce the velocity of runoff and check erosion. Increase in the yield of crops from 10 to 15 per cent has been reported as a results of contour farming. It is effective upto 2 per cent slope, on steeper slopes it ought to be supported by terracing and other measures.

Tillage

It is generally observed that a hard pan is formed due to continuous tillage at the same depth. Sometimes due to translocation of finner soil particles and their deposition at certain depth below the soil surface a hard impervious layer is formed. Percolation of water in such soils is poor. Deep tillage helps to break such formations which ultimately helps increase the infiltration rates and reduce the total volume of runoff and erosion of soil.

At the Soil Conservation Research Centre, Dehra Dun it has been observed that due to variations in number of ploughings for seed bed preparation no significant variations in the yield of maize was obtained. Similar results have also been obtained at other places also calls for the adoption of minimum tillage concept. For the seedbed preparation minimum amount of tillage sufficient to produce soil structure, conducive for crop growth should be practiced. Actually more soil manupulation cause destruction of soil structure so essential for maintaining the productivity of land. A well aggregated soil is less erosive.

Cover Crop

Actual measurements show that soil is susceptble to erosion when fields are bare of vegetation. Any crop while serving as a solid ground cover, whether or not specially planted for that purpose, is a cover crop. Thus a growing grain crop, grasses in pastures, and crops planted for turning under as green manure are cover crops. General usage, however, perhaps restricts the term more definitely to crops that are planted specially for the purpose of checking soil erosion, adding organic matter to the soil, and improving soil productivity. The advantages of cover crops is projected in the form of reduced runoff and erosion, increased infiltration of water, addition of organic matter to the soil, and improvement in soil tilth and improving the fertility status of the soil. But covercrop is not a cure-all. It represents only one of the recommended practices of a wall-rounded soil conservation programme.

Manures, Fertilizers and Soil Amendments

Additions of organic manures have been found to be helpful in increasing the amount of water stable aggregates in the soil which in turn reduces the volume of runoff and erosion of soil. Apart from cowdung, compost and green manuring, Lac mud, an industrial waste of lac industry has also been found to induce similar effects.

The use of fertilizers and amendments help luxuriant growth of crops providing better cover for the land which in turn reduces erosion. The benefits of traditional fertilizers and amendments have been well recognised which is manifested by their ever increasing demand. But in the recent years with the coming of new varieties deficiencies of micronutrients have been observed at many places. This needs proper attention for maintaining the productivity of soil at the desired level.

Crop rotation

Crop rotation helps in improving soil fertility and maintaining it at a high level. A good crop rotation involves the use of legumes and pulses alternated with cereals and other grain and cash crops. A well balanced crop rotation keeps the land almost continuously covered, change the location of feeding range of roots, improves the physico-chemical properties of soil. Reduces the bad effect of monocropping. Growing of soil depleting, clean tilled crops year after year depletes soil of its nutrients in the most unbalanced manner. Therefore, crops like soyabean, groundnut, mung, kalai, kulthi, cowpea, grasses, etc., which are fertility building crops should be incorporated in the general cropping pattern of the region.

Mixed Cropping

Experimental findings of D.V.C. have shown considerable reduction in soil loss as a result of growing maize or gora mixed with kalai as compared to growing sole crop of gora or maize. Thus it is better to grow mixture of legume with other crop instead of growing only one clean tilled crop. Birsa Agricultural University results have also shown that intercropping of legumes like arhar, groundnut, etc., with other crops is beneficial. Such practice helps build up the fertility status of the soil as well as reduce soil erosion.

Mulching

Tillage that leaves the surface of soil cloddy, minimise erosion and conserve moisture. That is why the farmers till their land in summer and keep in open to receive rainwater. Studies have shown that covering the land surface with the help of straw, stubbles, crop residues, paper etc., minimise the effect of raindrops impact and bring about considerable reduction in soil and water loss. Mulches after decomposition also add organic matter to the soil, increase infiltration rate of water and finally bring about improvement in the crop production. Central Arid Zone Research Institute, Jodhpur recorded an increase in grain yield of bajra upto 8.46 per cent as a result of stubble mulching.

Strip cropping

To reduce the velocity of running water more effectively strip-cropping method has been found to be very effective. It has been found to be effective in reducing not only soil and water loss but at the same time increase the productivity of land. Experimental results have shown an overall increase in the yield of crops to the tune of 40 to 100 per cent due to strip cropping.

This system involves the growing of alternate strips of erosion-permitting (clean tilled row crops) and erosion resisting crops (close growing cover crops), across the slope. If any soil is lost from the erosion permitting strip, if is deposited in the covercrop strip thus the soil is not lost from the field altogether.

There are four types of strip cropping methods :

(i) Contour strip - The strips are laid out strictly on contours.

(ii) Field strip - The strips are laid out approximately across the general field slope.

(iii) Wind strip - The strips are laid out across the prevailing wind direction against wind erosion.

(iv) Buffer strip - These are established to take care of the critical, vulnerable spots and permanently put under grasses and legumes.

Based on results of different research centres the following strip width may be recommended for this area for different slopes.

	Width of strip in metres	
Slop %	*Erosion permitting crop.*	*Erosion resisting crop.*
1%	50	10
2%	24	6
3%	15	5
1%	24	4
2%	24	8
3%	24	10

Forestry and Pasture Management

Both forestry and grassland development are essential for the progress of agriculture, as they influence productivity in some way or other by their direct or indirect effects. Therefore land which are nor suitable for cultivation should invariable be put under some forest crop or grasses. For this locality tree species like sal, teak, sishum, bamboo, kathal, mango, eucalyptus, gamhar, etc., have been found to be well suited. It is economical too because a timber tree after maturity say after 20 years or so will be worth substantial value.

There is great need for firewood which can be mét by planting quick growing trees like different species of acacia, albazia, cassia, etc.

Many a times there is shortage of green fodder for the cattle. This can be tide over by growing such forest crops like subabul green leaves of other trees will also provide feed in time of scarcity.

The status of grasslands and pastures are generally very poor in our country due to unscientific management. This position has to be changed by introduction of good palatable varieties and scientific management practices.

We will not be going into full details of these measures but few common tips are being mentioned here.

For establishment and upgrading the forests or pasture the first and foremost this is to fence the area to avoid stray cattle. The common fencing method is barbed wire, trench fencing, stone wall fencing etc. After fencing staggered contour trenches are dug to help soil and moisture conservation. Thereafter planting and seeding is done.

Rational method of timber extraction and pasture management should be followed. Farm forestry should be encouraged and growing of fruit trees may be incorporated along with timber crops to enhance income of farmers.

Shifting cultivation practiced at some places cannot be checked all of a sudden as it is practiced as a way of life. But some system of improvement may be introduced to reduce, its harmful effect. It may be made obligatory to afforest the area before shifting to another land and payments for such afforestation works may be done by Government to provide incentive and make the procedure full proof. Apart from this till the trees grow up to the vacant land in between can be put under cultivation, so that the land is not completely open when it is being left.

The banks of streams and rivers must be put under forests. An eminent soil conservationist had opined that if 100 meters wide forest is raised all along the river or stream course havoc of flood, draught, stream bank erosion etc., can be reduced to a minimum.

The pastures land are invariably most neglected. For upgrading such lands, fencing the area to stop stray cattle and allow it for natural regeneration is essential. Apart from this weeding of good palatable varieties of grasses and legume; like cenchrus, dicanthium, cynodon, panicum, kidru etc., should be done. Shallow staggered trenches across the slope, rotational grazing have been found to be beneficial from the point of view of increasing production and reducing soil and water losses.

Engineering Measures

Though the cheapest method of conserving soil would be through agronomic practices, it is not possible to employ only these practices for the success of such programmes. Employing mechanical measures becomes a necessity.

Terracing

The important principles involved in the design of terraces are : change the general feature of the land slopes, reduce the total volume of runoff and arrange for the safe disposal of excess runoff.

For this region the following types of terraces have been found suitable.

(i) Bench terrace

It is perhaps the earliest conception of terracing. The practice of bench terracing, the low lying fields for paddy cultivation, is an age old practice of this region. The practice is to construct a series of level platforms at suitable intervals by cutting and filling of each, and having a bund at the outerside to arrest runoff.

Bench terraces, level or graded longitudinally, are the following three types :

(a) Level or table top.

(b) Sloping outward.

(c) Sloping inward.

But for this region table top or level bench terrace has been found to be most suitable.

For calculating the width of individual fields the general formula is as given below :-

$$W = D\,\frac{(100-S)}{S}$$

Where W = Width of bench terraced field in ft.

D = vertical interval between terraces in ft.

S = per cent slope of land.

(ii) Level or absorption types terraces

Narrow based bunds of suitable cross section from 6 to 8 sq. ft. is constructed on contours. Their function being to intercept and allow absorption of water by the soil. This type of terrace is not vary in this locality.

(iii) Graded terrace

Narrow based bunds and channels of 8 sq. ft. cross section with uniform grade of 0.1 per cent has been found to be most suitable for this region. The grade is increased to 0.25 per cent in the last 100 ft length of terrace for better disposal of excess water.

The general formula for calculating the vertical interval of terrace is:

$$VI = \frac{S}{2} + 2$$

Where VI = vertical interval in feet

S = per cent slope of land.

The length of slope in one direction should not normally exceed 600 to 800 ft due to poor maintenance of structures in the field.

(iv) Tati terrace

This type of terrace was first adopted in the village Tati of Palamau district by D.V.C. authorities and so named after it.

In this method the plots are closed from all sides by bunds of suitable cross section generally varying from 4 to 6 sq. ft., leaving suitable spillway on one. The ultimate objective is to have a level field without much cutting and filling of earth. The only technical point kept in view in this that the diagonal fall does not exceed two fact. This is a compromise to the cultivators dislike for the grade terrace in which their fields are split into plots of unconventional shape and size.

(v) Conservation Terrace

It will not be out of place to mention here about the "Conservation Terrace" designed by D.V.C. people, instead of having graded bunds as such, what they have done is this that the lower 1/4th to 1/5th part of the terraced plot along the bund in the lower portion has been made level and one foot lower than the general ground surface. All the water from the upper portion of the plot is made to collect in this dugout portion. The water remains in it as it is closed at both ends. This water is utilized for raising paddy crop. The data shows that it has given good results specially in medium lands with heavier soils. I have seen this type of terrace of Deochanda Research Station, Hazaribagh (D.V.C.). But it has not come out of the boundaries of Deochanda farm. This needs actual testing in the farmers fields. I feel that as this involve the principle of watershed and water-harvesting, it may prove a good break through in this direction.

Results of Deochanda (Hajaribagh), Rahmankhoa (U.P.) have shown that terracing brings about a marked improvement in the productivity of land.

In a systematic study undertaken at the Ranchi Agril. College it was revealed that as a result of the construction of both tati and graded terraces marked improvement in the physio-chemical properties of soil and productivity of land has taken place. The crop yield showed an average increase of 20 to 75 per cent. The analysis of cost benefit ratio, based upon 10 years beneficial life of terraces and calculation of depreciation and interest by the method of "Equated system of loan recovery (Principal + Interest)" followed by Land Mortgage Bank, showed an average increased return of 22 to 30 times the investment.

Grassed water ways

In terracing projects sometimes grassed waterways are also constructed for safe disposal of runoff water. Generally it is constructed in trapizoidal shape which takes the shape of parabola in due course of time. The side slope is generally kept much flatter 2:1 to 4:1. Velocity of flows should not exceed 1 to 5 m/sec. Using this velocity the carrying capacity of the waterways is calculated by formula :

Q = av, where Q = Discharge

a = Cross sectional area

v = Velocity.

The design is based on peak discharge which is expected from the area and this is estimated by rational formula.

Q = CIA, where Q = Peak discharge in Cu. ft/Sec.

C = Constant depending upon the watershed characteristics.

A = Area of watershed in acres.

I = Intensity of rainfall in inches per hour for a duration equal to the time of concentration.

The grassed waterways is generally constructed a year ahead of the construction of terraces, so that grass established well before the water is turned into it. Afterwards it should be properly maintained. Any neglectful act may results in transforming the waterways into a gully.

Gully control and reclamation

Gullies are a symptom of a functional disorder of the land as a result of improper landuse. But gullies are not the first symptom of land misuse, they are generally a secondary symptom, indicating misuse of land over a long period of time.

The most important work in gully reclamation is to reduce the runoff through adopting water conservation measures in its watershed. In case of small gullies and planting some vegetation is essential to stabilise it.

In case of active gullies the main work will be to divert the runoff through diversion channels to avoid further damage. Gullies develop in three directions elongate along head, expand in width and increases in depth.

Thus control works are also of three types :-

(i) Gully side control

(ii) Gully head control

(iii) Gully bed control.

(i) Gully side control

Periferal bunds should be constructed to check the entry of water from the sides. The side slopes should be eased out and vegetation should be established..

(ii) Gully head control

If the drop is small sod flume will serve the purpose.

In case the drop is more or volume of incoming water is more sod flumes will not work and masonry structures, are required. In such cases Chute spillway of suitable size is required to be constructed. This will safely lead the water from the higher level to the bottom of the gully.

(iii) Gully bed control

For stabilization of the gully bed construction of check dams may be needed at suitable places to reduce the velocity and force of moving water and harvest the soil coming with the water. These structures may be temporary in case of small gullies or permanent in case of bigger ones. The temporary check dams can be constructed with any locally available materials like rock, timber brushwood, wiremesh etc., Along with this some sort of vegetation may be established.

In case of bigger, active gullies and where the intake of runoff is more in place of temporary structures, permanent structures like earthen dams with suitable masonry spillways are to be constructed. If the drop is less than 10 ft. and there is less chance of deposition of silt pipe spill way or drop spillway may be the choice. But if the drop exceeds 10 ft and there is chances of accumulation of silt then drop inlet spillways should be preferred. Apart from this gully bed should be levelled and vegetation should be established.

In case of considerably wider gullies the beds may be levelled and put under cultivation if possible.

Silt detension cum water harvesting dams

At suitable sites where the grass water ways end or there is accumulation of water from the project area silt detension cum water harvesting dams are also constructed at arrest whatever silt is being washed out and to the same time harvest the excess runoff for being utilized in raising crop after rainy season.

In the construction of such dams proper design of spillway and corewall is of prime importance. The spillway should be designed at proper place, on the basis of peak discharge rate, keeping in view to give good storage capacity to the reservoir. Proper construction of corewall is necessary to check seepage losses which is quite much in earthen dams. Corewall should invariable be made of clay material. Sides of the earthen dams should not be given proper slope depending upon the soil type for its stability. For average quality of soil material it should not be steeper the 3:1 on the upstream side and 2:1 on the downstream. Coarse soils may need side slopes of 3:1 to 4:1. The upstream side should be protected with riprap against wave action, whereas down stream should have good turf for protection. The height of the dam should be determined after fixing the height of spillway. The dam height should be at least 2 ft above the spillway height. The top width may be kept anything between 8 to 12 as per requirements.

Spurs and Levees

Spurs are constructed as projections from a bank of a stream so as to deflact the flow of water and check stream bank erosion. The head of spur must be well built into the natural bank of the stream and placed at some angle down stream and at interval all along the vulnerable spot. It may be temporary, permeable of boulders, brushwood etc., or permanent of masonry work.

Levees are earthen dams constructed along the flow on the banks of the stream to protect the adjoining areas as from flood and check stream from changing its course. The inside wall of levees are kept quite flat 6:1 to 8:1 and protected by rip rap. The outside wall is kept a bit steeper 2:1 to 4:1 give stability to the structure and protected by sodding.

Chapter 13

Role of Mycorrhiza to Improve Yield in Agroforestry

*N.K. Verma**

Agrosilviculture, silvipastoral, agrosilvipastoral systems or home garden etc. are the different forms of agroforestry systems which have been in practice world wide since long. In recent past, the agroforestry has assumed prominence in the wake of the increasing population pressure, consequent destruction and mismanagement of forests by man is the quest of food and wood products and the resultant environmental problems. Increasing dependence of modern agriculture technology on high value inputs on the one hand and the deteriorating economic situation of the most developing countries on the other have caused a renewed awareness about the productive and protective value of trees and realisation of the potentials of age old conservation farming technology.

The general objective of agroforestry research is to uncover the relationship between the performance of tree and non-tree mixtures, particularly their productivity and sustainability in response to climate, weather, management and characteristics of the component species. As the two components, one tall woody perennial trees and other the small herbaceous crops are raised on the same piece of land in agroforestry system, it involves a complex micro-climatic interaction between the two. For continued growth, plants require continuous and balanced access to resource pools of light, water and nutrients. Except light the other two resources gradually diminish replenish or lost from the system in cases in response to growth. The microbs and symbiotic phizobiol, actinorrhizal and mycorrhizal activity in the soil and roots are directly related to nutrient release in soil and their availability to plants. The excessive use of fertilizers, insecticides and fungicides in order to grow more and more has virtually toxicated most of the farm soils, consequently, reducing the

---

**Faculty of Forestry B.A.U., Ranchi (Bihar)*

natural microbiol activity in soil. This agriculture practice coupled with the constant soil erosion problem particularly in undulating and hilly areas has resulted into a very poor fertility status of the farm land. The reduction in natural microbiol population, the vesicular arbuscular mycorrhizal fungal (VAM) pore population and rate of nitrification in the disturbed lands (Sharma *et al.* 1982) has the potential to initiate the process of desertification. The complete washing out of the ectomycorrhizal propagules from the soil has resulted into the elemination of the native Khasi pine (pinus kesiya) from Cherrapunjee (Verma and Shukla, 1989).

Keeping the land fallow for few years after growing the crops in the land was a very scientific age old practice to allow the soil naturally regain the fertility in the natural way. Because of the population pressure this practice could not and should not continue. Alternatively, the option left may be to artificially treat the land to accelerate the microbiol and symbiotic activity to regain the fertility of the soil. This paper reviews the possibility of increasing the growth and yield of the agricultural and forestry crop on the same piece of land by introduction of vesicular-arbuscular mycorrhiza (VAM).

Microbial and symbiotic activity in soil

The surface soil layer is considered as the zone of intensive root activity with the sub-soil constituting the extensive root activity zone. In the ecological (Mosse, 1973, Bowen, 1980), forestry (Harley, 1970) and agronomic literature (Khan, 1975, Hale and Armstrong 1977, Asimi *et al.* 1980) the outstanding feature of nutrient uptake is the range of symbiotic relationship with the micro-organisms that plants have developed over the centuries to improve nutrient uptake in nutrient poor soil. Probably, the very evolution of land plants was possible only through such mutualistic partnerships that were equipped to cope with the problems of desiccation and starvation associated with the terrestrial existence (Pirozynski and Malloch, 1975). While rhizobial associations are restricted to the legumes and a few others, mycorrhizal associations of one type or other are almost universal in the roots of natural vegetation and can be found in most agricultural crops and pastures. Mycorrhizal associations are known to increase the yield of the host plants under conditions of low fertility and the nutrients most firmly implicated is P, although nutritional advantage with respect to N, Ca, Na, Mg, Fe, Mn, Cu, B and Al - have also been recorded (Mosse, 1973).

The nitrogen fixing rhizobial symbiotic association and that of the mycorrhizal symbiotic associations have important differences with respect to overall nutrient relationships. Rhizobial associations place at disposal of the symbionts the enormous reserve of gaseous nitrogen which is unavailable directly to non-nodulated plants. By contrast mycorrhizal associations do not allow access to proviously unavailable reserves of nutrients. The fungal hyphac which ramify extensively through the soil from infected roots absorb P from the same nutrient pool as uninfected roots. The major advantage appcars to be the additional exploration potential of infected plants (Sanders and Tinker, 1973). But the advantage of symbiotic bacterial nitrogen fixation is confined to "elite" group of plants i.e. legumes whereas the advantage of mycorrhizal P uptake is widely distributed throughout the plant kingdom. Even the legumes do require certain levels of phosphorus for growth and effective nodulation (McLachlan and Norman, 1961, Gibson, 1976) and tropical legumes are much more dependent on mycorrhiza for growth than the temperate species (Bagyaraj, 1984), Nosse *et al.* (1976) showed that effective nodulation of legumes in very P-deficient Brazilian Cerrado soil could be achieved only by introducing both mycorrhiza and extra P (rock phosphate). As tropical soils are very often deficient in phosphorus and P fertilizer is expensive, it is one of the bottlenecks of tropical agriculture/forestry/ agroforestry and even slight improvement in mycorrhizal efficiency might be of great practical importance (Mikola, 1980).

VAM in Agriculture

The report on growth response to VAM inoculation in agricultural crops is not in plenty. Among the non-legumes much work has been done on barley. 290 per cent increase in grain yield due to VAM inoculation along with the increase in vegetative growth in barley was the first ever report on field trial (Saif and Khan, 1977), although it did not give any statistical analyses. Remarkable increase in yield and vegetative growth has been reported with VAM in barley (Owusu-Bennoah and Mosse, 1979), Powell et al. (1980). Khan (1972, 1975) in his field trial reported very long growth responses in shoot dry matter and grain yield to VA-mycorrhizal inoculation in case of maize and wheat on transplant from nursery beds to the farmer's field. Govind Rao *et al.* (1983) found that mycorrhizal inoculation of finger millet with a selected VAM strain (M6) gave 18 per cent better grain yield than non-inoculated plants over three fertilizer rates. Similar positive response with respect to yield, growth rate and nutrient absorption has been found in the crops such as Cassava (Yost and Fox, 1979, Van der Zaag *et al.* 1979, Onion Lowusu - Bennoah and Mosse, 1979, Powell and Bagyaraj, 1982, Mishra and Verma, 1983), Potato (Black and Tinker, 1977, Kormanik *et al.* 1980), Sunflower (Iqbal and Qureshi, 1977) and Chili (Bagyaraj and Sreeramulu, 1982).

Among the legumes, most of the reported filed trials are available with the crop Soyabean. Ross and Hasper (1970) inoculated Soybeans in sterilized and unsterilized soil in Mississippi and found a 29 per cent response in grain yield to VAM inoculation in sterilized Soil. Schenck and Hinson (1973) transplanted Soybean seedlings with or without *Gigaspora gigantea* inoculation into a sterilized field soil and found a 53 per cent respose in grain yield of a nodulating isoline but no response to inoculation of a non-nodulating isoline. Increased yield and drymatter of soybean after VAM inoculation has also been reported in unsterilized field soils (Kno and Huang, 1982, Bagyaraj *et al.* 1979). Positive response to VAM inoculation has been reported in cowpea and lucerne also.

VAM in woody perennials

About the ubiquity of mycorrhizal associations with tree Kormanik (1984) has rightly said "I cannot say that every tree in the forest has mycorrhiza but I can say that every tree I have ever examined for mycorrhiza had them". Agroforestry on agricultural land essentially requires the plantation of tree species (including fruit, fodder, fuel and timber species), and in plantation crop the scope of VAM inoculation has far better prospect compared to natural forest regeneration. Powell (1984) has summerised the performa of fruit, timber and ornamental trees with VAM association. VAM inoculation not only increased the growth of citrus but also helped to overcome the stunting effects of sterilization on plant vigour (Kleinscmidt and Gerdemann, 1972, Martin, 1948, Martin *et al.* 1953 on the field performance of VAM inoculation nursery stock, Schinck and Tucker (1974) observed that sour orange had 34 per cent more shoot length in disked unfunigated soil than in equivalent control plants at 100 days after transplanting Minge *et al.* (1980) increased the growth of avocado seedlings by 49 to 254 per cent in sterilized loamy sand by inoculation with two isolates of *Glomus fasciculatum* Mege *et al*, (1978) also observed decreased transplant shock in mycorrhizal compared to non-mycorrhizal avocado plants. 79 per cent increase in the height of peach seedlings inoculated with *G.fasciculatum* in sterilized nursery soil was observed by La Rue *et al.* (1975), Plenchette *et al.* (1981). Observed 142 per cent increase in shoot length in unsterilized field soil through VAM inoculation. Among the hardwoods, Kormanik *et al.* (1977), found the sweetgum, almost obligate mycorrhizal for good growth over non-mycorrhizal plants. Mycorrhizal growth response of 43 to 167 per cent was observed (Kormanik *et al.*, 1977) in different clonal lines of yellow poplar which decreased the time required for plants to reach a saleable size. (Mukherjee and Jagpal, 1987) observed that VAM inoculation *Leucaena leucocephala* not only showed better growth performance but also increased the vigour of the plant to overcome the adverse conditions such as salinity, low fertility and drought in wastelands near Delhi *L. leucocephala* has been shown to have very high dependence on VAM (Habte and Manjunath, 1987).

VAM in Agroforestry

Either the forest land or the agricultural land is used for agroforestry. When the land use pattern is changed it is un-predicatable to expect the behaviour of VAM. The incidence of mycorrhizal fungi changes in quantity and quality with ecosystems (Schenck and Sequiera, 1987). In general, there is not much host specificity in VAM but when a tree and agricultural crop is grown together, there is bound to be a change in VAM species composition. For greater commercial point of view, the most effective VAM fungal species has to be screened out, but it requires area specific research on this line. Since the species incidence of VAM mycorrhizal fungi changes with ecosystem and is effected by levels of soil variables (pH, soil fertility, soil types, host effects and other items) we need to be aware of the ability of a species to persist in any particular ecosystem or fertility level. This is specially true when we wish to use that species in a field inoculation programme.

Conclusion

Agroforestry system may be said to be the ideal system of introducing the superior strain of VAM. Because of non-host specificity in VAM, the introduction of VAM may be beneficial to agricultural and forestry both the crops. The ideal situation of VAM introducing may be the nursery stage. The common nursery practice in India and other tropical and sub-tropical countries is to raise the seedlings in various kinds of containers such a clay pots, wooden boxes or polythene tubes. It is at this stage the VAM can be lasily inoculated in the seedlings when these VAM inoculated seedlings are planted in field, the VAM may spread in soil through root to root contact. If the nitrogen fixing trees are also introduced along with VAM it may enrich the soil with N and P through natural biological activity in soil. However the experimental trials on these lines are required before finally recommending it for the mass scale operations.

REFERENCES

Asimis, S., Gianinazzi-pearson, V. and Gianinazzi, S. (1980). Influence of increasing soil phosphorus levels on interactions between Vesicular-arbuscular-mycorrhizal and *Rhizobium* in soyabeans, *Can. J. Bot., 58*, pp. 2200-5

Bagyaraj, D.J., Manjunath, A., and Patil, R.B. (1979). Interaction between a vesicular-arbuscular mycorrhiza and Rhizobium and their effects on Soyabean in the field, *New Phytol., 82*, p. 141

Bagyaraj, D.J., and Sreeramulu, K.R. (1982) Pre-inoculation with VA mycorrhiza improves growth and yield of chilli transplanted in the field and saves Phosphatic fertilizer, *Plant Soil, 69*, 375.

Bagyaraj, D.J. (1984): Biological interactions with VA Mycorrhizal fungi, In : VA mycorrhiza (Eds. Powell. C.L. and Bagyaraj, D.J.), J.C.R. C. Press Inc. Boca Raton, Florida, pp. 131-153.

Black, R.L.B. and Tinker, P.B., (1977). Interaction between effects of Vesicular-arbuscular mycorrhiza and fertilizer phosphorus on yields of potatoes in the field, *Nature*, (London), pp. 267, 510.

Bowen, G.B. (1980). Mycorrhizal roles in tropical plants, *Ann. Rev. Ecol. Syst.*, II, p. 233

Gibsson, A.H. (1976): Limitations to nitrogen fixation in legumes, In : *Proc. Ist. Int. Symp. Nitrogen fixation*, Vol. 2 (Ed. Newton, W.E. and Nayman C.J.) Washington State Univ. Press. Pullman, p. 410.

Govind Rao, Y.S., Bagyaraj, D.J. and Rai, P.V. (1983). Selection of an efficient VA mycorrhizal fungus for finger millet II. Screening under field conditions, *Zentralbl. Microbiol.*, p. 138, 415.

Habte, M. and Manjunath, A. (1987). Soil solution phosphorus status and mycorrhizal dependency in *Leucaena leucocephala, Appl. and Env. Micro.*, 53(4) 797.

Hale, I.R. and Armstrong, P. (1979). Effect of Vesicular-arbuscular mycorrhizas on growth of white clover, lotus and rye grass in some eroded soils, *N.Z.J. Agric. Res.* 22, 479.

Harley, J.L. (1970). Mycorrhiza and nutrient uptake in forest trees. In *"Physiology of Tree crops and cutting,"* eds : Luckwell, L.C. and Cutting, (Acad, Press, London.), pp. 163-79.

Iqbal, S.H. and Qureshi, K.S. (1977). The effect of Vesicular-arbuscular mycorrhizal association on growth of Sun flower (*Helianthus annus* L.), under field conditions, *Biologia* (Lahore), 23, p. 189.

Khan, A.G. (1972). The effect of Vesicular-arbuscular mycorrhizal associations on growth of cereals I. Effects on maize growth, *New Phytol.*, II, p. 613.

Khan, A.G. (1975). The effect of Vesicular-arbuscular mycorrhizal associations on growth of Cereals II. Effect on wheat growth, *Ann. Appl. Biol. 80*, p. 27.

Keinschmidt, G.D. and Gerdemann, J.W. (1972). Stunting of Citrus seedlings in fumigated nursery soils related to the absence of endomycorrhizal. Phytopathology, *62*, p. 1447.

Kormanik, P. (1984). Applications of VA mycorrhizal fungi in Forest tree crops, In : *Applications of mycorrhizal fungi in crops production* (Ed. Ferguson, J.J.) Univ. Florida, Gainseville, U.S.A. p. 48.

Kormanik, P.P. Bryan, W.C. and Schultz, R.C. (1977). Influence of endomycorrhiza on growth of sweetgum seedlings from eight mother trees, For Soi. 23, p. 500

Kormanik P.P., Bryan, W.C. and Scultz, R.C. (1980). Increasing endomycorrhizal fungus inoculum in forest nursery soil with cover crops, South, *J. Appl. For.*, *4*, p. 151.

Kormanik, P.P. Bryan, W.C. and Scultz. R.C. (1977). Endomycorrhizal inoculation during transplanting improves growth of Vegetatively propagated yellow Poplar, *Plant. Prop. 23*, p. 4

Kuo, C.G. and Huang, R.S. (1982). Effect of Vesicular-arbuscular mycorrhizal on the growth and yield of rice stubble cultured soybeans, *plant soil*, *61*, pp. 325.

La Rue, J.H. Meclellan, W.D. and Peacock, W.L., (1975). Mycorrhizal fungi and Peach nursery nutrition, *Calif. Agric.*, *29*, p. 6.

Martin, J.P. (1948). Effect of fumigation, fertilization and various other soil treatments on growth of orange seedlings in Old Citrus soils, *Soils Sci. 66*, p. 273.

Martin, J.P. Aldrich, D.G., Murphy, W.S., and Bradford, G.R. (1953). Effect of soil fumigation growth and chemical of soil fumigation on growth and Chemical composition of Citrus plants, *Soil Sci.* 75, p. 137.

McLachlan, K.D. and Norman, B.W. (1961). Phosphorus and Symbiotic nitrogen fixation in subterranean clover, *J. Aust. Inst. Agric. Sci., 27*, p. 244

Menge, J.A., Davis, R.M. Johnson, E.L.V., and Zentonyer, G.A. (1978). Mycorrhizal fungi increase growth and reduce transplant injury in avocado, *Calif. Agric. 32*, p. 6.

Menge, J.A. La Rue, J., Lananauska, C.K. and Johnson, E.L.V. (1980). The effect of two mycorrhizal fungi upon growth and nutrition of avocado seedlings grown with six fertilizers treatments, *J. Am. Soc. Hortic. Sci. 105*, p. 400.

Mishra, R.R. and Verma, N.K. (1983). Effect of different mycorrhizal treatments on the growth of Onion, *Acta. Bot. Indica., 11*, p. 49.

Mikola, P. (1980). Mycorrhizae across the frontiers. In. *Tropical Mycorrhiz Research* (Ed. Mikola, P.) Clarendon Press, Oxford, p. 3-4.

Mosse, B. (1973). Advances in the study of Vesicular-arbuscular mycorrhizal, *Ann. Rev. Phytopathol., II*, p. 171.

Mosse, B., Powell, C.L. and Hayman, D.S. (1976). Plant growth responses to vesicular-arbuscular mycorrhiza IX. Interactions between V.A. mycorrhiza rock Phosphate and symbiotic nitrogen fixation, *New Phytol., 76*, p. 331.

Mukherji, K.G. and Jagpal, R. (1987). VAM in reforestation in Indian waste lands. *7th North. Amer. Conf., Mycorrhizae.* p. 161.

Owusu-Bennoah, E. and Mosse, B. (1979). Plant growth responses to vesicular-arbuscular mycorrhiza XI. Field inoculation responses in barley, lucerne and Onion, *New Phytol. 83*, p. 671.

Pirozynski, K.A. and Malloch, D.W. (1975). The Original of land plants a matter of mycotrophism, *Biosystems,* p. 6.

Plenchette, C. Furlen, V., and Fortin, J.A., Growth (1981). Stimulation of apple trees in unsterilized soil under field conditions with VA mycorrhizal inoculation,. *Can. J. Bot., 59*, p. 2003.

Powell, C.L. (1984) : Field inoculation with VA mycorrhizal fungi, In . *VA mycorrhiza* (Eds. Povell) (Raton Florida,) p. 131.

Powell, C.L. Grotrers, M. and Metcalfe, D. (1980). Mycorrhizal inoculation of a harley crop in the field, *N.Z.J. Agric. Res.*, 23, p. 107

Powell, C.L. and Bagyaraj, D.J. (1982). VA mycorrhizal inoculation of field crops, *Proc. N.Z. Agron. Soc., 12*, p. 85

Ross, J.P. and Harper, J.A. (1970). Effect of Endogone Mycorrhia on Soybean yields, *Phytopathology, 60*, p. 1552.

Safi, S.R. and Khan, A.G. (1977). The effect of Vesicular-arbuscular mycorrhizal associations on growth of Cereals, III. Effects on barley, growth, *Pl. Soil, 47*, p. 17.

Sanders, F.S. and Tinker, P.B. (1973). Phosphat flow in to mycorrhizal roots, *Pestic. Sci., 4*, p. 385

Schenck, N.C. and Hinson, K., (1973). Response of nodulating and non-nodulating Soybeans to a species of Endogone mycorrhiz, *Agron. J.* 65, p. 849.

Schenck, N.C. and Tucker, D.P.H. (1974). Endomycorrhizal fungi and development of Citrus seedlings in florida fungigated soils, *J. Am. Soc. Hort. Sci., 99*, p. 284.

Schenck, N.C. and Sequeira, J.O. (1987). Ecology of VA - mycorrhizal fungi in temperate agroeco-systems. *Proc. 7th North Am. Conf. Mycorrhizae*, Florida, U.S.A. p. 2.

Sharma, G.D., Tiwari, B.K. Saxena, K.G. Verma, N.K. and Mishra, R.R. (1982). Mycorrhizal and Soil microbial Characters as influenced by shifting agriculture. Training Courses on mycorrhizal research techniques. *Int. Found. Sci. Sweden*, p. 408.

Verma, N.K. and Shukla, R.P. (1989). Ectomycorrhizal status and growth of Khasipine (*Pinus kesiya*) in the soils of disturbed land of Cherrapunjee, *Indian forest, 115*, p. 337.

Vander Zaag. P. Fox. R.L., de la Pena, R.S. and Yost, R.S., (1979). P nutrition of Cassava, including mycorrhizal effects on P.K.S. Zn and Ca uptake, *Field crops Res., 2*, p. 253.

Yost, R.S. and Fox, R.L. (1979). Contribution of mycorrhizae to P nutrition of crops growing on an oxisol, *Agron. J., 71*, p. 903.

Chapter 14

Palynology in the Genetic Improvement of Trees Suitable for Agroforestry

*C.M. Sharma**

Introduction

It is conceivable that tree genes migrate via pollen or seed. Pollen transfer is gamete transfer and seed transfer is zygote transfer. Immigration of pollen into a hybrid means immigration of genes into the hybrid. Migration of the genes is a one way process. Different pollen species have different migration rate. If the forest stands are dense or if the pollen dispersion distances are great the gene flow is also rapid, the entire population thus remains uniform. As contrast if the population is sparse or if the pollen dispersion distances are small, gene flow is inhibited and the different segments of the population can become differentiated genetically.

The important aspects for tree improvement may be selection and crossing of trees from different geographical origins and production of hybrids with the help of the exotics, polyploidy breeding etc. The pollen grains are involved in each breeding process of tree improvement programmes. Taking it into the consideration, some important methods are discussed briefly, which can be of immense help in tree genetic improvement programmes.

Pollen in Tree Breeding

The influence of male parent is manifested in F_1 generation as a rule, provided the character is dominant, but in some cases the pollen may produce a direct effect on the parts of seed and fruit (Xenia and

**Department of Forestry, HNB. Garhwal University Srinagar Garhwal-246174 (U.P.)*

Metaxenia). Various effects like shape and size of both fruit and seed, number of seeds per fruit, yield, quality, colour and ripening time of the fruits, have been found to be influenced by foreign pollen. Some important roles of pollen in tree breeding can be summarized as follows:-

Pollen grains are considered to be the fittest material for irradiation studies. Brewbaker and Emery (1962) have shown that irradiation sensitivity is related to pollen size. The effect of irradiation is sometimes also reported upto M-2 generation. The effect is more striking in polyploid materials since polyploid gametes can withstand chromosome aberration to a greater extent, than that of diploids (Swaminathan, 1964). Induction of mutation through X-Ray irradiation of flower buds at pollen mother cell stage was one of the earliest successful techniques for the induction of self-incompatibility (Lewis, 1949).

By hybridization the crossing of diverse species is done to achieve genetically superior progeny. In forestry crossing of different species is done to enhance the growth rate better than that found in either parent. Combination of favourable traits from two or more different species is another practical goal of hybridization work. Knowledge of natural distribution of the species is also essential for crossability pattern. Repetition of successful crosses, testing and selection leads to genetically superior cross for better hybrids.

Pollen behaviour

In most species a high percentage of trees do not flower regularly, or, having flowered, they set small amounts of seeds. In controlled pollination artificial transfer of pollen is done from one flower to another, and male and female parent are known with certainity. Usually artificial transfer of pollen grain is possible only when the female flowers are enclosed in bags well before pollination time. Controlled pollination studies are necessary parts of species hybridization or racial hybridization and selective breeding programmes (Larsen, 1956).

In insect pollinated species pollen is scarce and must be applied by brushing on to each stigma individually. In experimental programmes different pollens are applied on the same tree. This necessitates separate labelling of each pollinated branch.

Pollen dispersion distance is more critical in most tree genetic work. Small amount of pollen travel hundreds of miles from the nearest land (Erdtman, 1954).

Pollen Germination and Tube Growth

In any breeding programme, pollen germination studies are essential for estimation of the quantity of the pollen required for controlled pollination. This information is lacking with regard to the Indian hardwoods as only a few records are available. However, pollen germination and viability of six hardwoods i.e., *Juglans regia, Rhododendron arboreum, Pyrus pashia, Populus deltoides, Populus ciliata* and *Salix tetrasperma* have been undertaken by Khurana and Khosla (1969), and some other trees by Sharma and Gaur (1984), Gaur and Sharma (1985), and by many other workers time to time.

The pistil provides the pollen with an environment, which is most fit for the pollen germination and tube growth *in vivo*. But certain growth regulators like, boron, hormones, carbohydrates (sugars), antibiotics, extract from flowers, and irradiation are known to have stimulating effects on pollen germination and tube growth *in vitro* (Dugger, 1973; Gaur and Sharma, 1985; Sharma 1984; Brewbaker and Emery, 1962; Linskans, 1974 etc.), alongwith moisture, CO_2 and O_2 levels, temperature, pH, osmotic pressure and certain other minerals, *in vivo* as well as *in vitro*. The germination requirements of pollen grains differ much from species to species.

Pollen Viability and Longevity

Environmental factors, particularly temperature and humidity, greatly effect the pollen viability. Retention of pollen viability after shedding, varies significantly from species to species. There is a close correlation between the cytology of pollen (two or three celled) and loss of viability. Two celled pollen generally retain viability for a longer period, then three celled pollen (Brewbaker, 1957). Successful pollen storage is a very convenient tool in the hands of tree breeders for improving the trees by hybridization, occurring in different regions and also of those blooming in different seasons. Apart from pollen use in self-incompatible groups of trees, stored pollen are also used in repeating a particular cross in subsequent years with the same pollen. The longevity of pollen grains enables the introduction of new characters in trees by using stored living pollen. Thus there are many uses of pollen storage in tree improvement programmes. The trees breeders have therefore, established 'Pollen banks' through which the viable pollen of the desired species can be obtained at any time. By this contrivance, supplementary pollinations are done for genetic improvement of different trees.

Pollen Sterility and Incompatibility

The knowledge of pollen sterility and compatibility is more important to understand the pollination mechanism and breeding behaviour of the species. Selective male sterility has important practical applications and extensive work has been carried out, particularly on cytoplasmic male sterility (Edwardson, 1970; Laser and Lersten, 1972). Cytoplasmic male sterility is inherited through the female parent.

Hybrid seeds can be produced in male sterile lines by crossing them in controlled pollination conditions. Commercial production of hybrid seeds essentially depends on the prevention of self or intraline pollination in individuals used as female parent, and effective transfer of pollen from the male parent to stigmas of female parent. The use of male sterile lines has revolutionized hybrid seed production programmes as contrast to the convention methods like emasculation, bagging and manual pollination, which are too laborious. The mechanism of selfsterility has been worked by several workers i.e., *Pinus* (Johnson, 1945), *Castanea* (Mc Kay, 1942), *Picea* (Langner, 1959), *Eucaltus* (Pryor, 1957), *Pseudotsuga* (Orreweing, 1957) and many others.

Self-incompatibility is one of the most effective outbreeding mechanisms, and is considered to be one of the main causes, for the rapid evolution of angiosperms. Besides its significance in reproductive biology of angiosperms, self-incompatibility has many other practical applications (Reimannphillipp, 1965).

Pollen Chemistry and Energetics

Variation in chemical constituents of pollen is accounted for by (i) Species differences, and (ii) environmental differences during maturation and after dehiscence. During maturation excessively high temperatures tend to reduce the pollen carbohydrates, low light also results in less carbohydrates accumulating in mature pollen, probably as a direct effect of reduced photosynthesis. If the plant nutrient supply, particularly microelements, is below optimum, pollen mineral content may be reduced. Lower content of trace-elements in pollen can modify, the protein-enzyme levels and thus effect growth and seed set of the tree (Latzko, 1964).

Wind-pollinated trees irrespective of their taxonomic position, clearly have lower pollen caloric contents than insect pollinated species (Petanidou and Vokou, 1990). The wind pollinated species produce, in general, pollen grains rich in carbohydrates (Baker and Baker, 1979). As contrast, insect pollinated species are frequently lipid rich. As lipids contain much more energy per gram than starches, lipid rich pollen would have higher energy content per unit of mass. The higher amounts of energy are required if pollen tubes have to be traverse long styles. It could be concluded, therefore, that higher energy contents of pollen from insect pollinated plants would be the result of selective pressure to secure transport of the male gametes from the anther to the ovule. It would mean that pollen energy content could increase attractiveness to pollen consumers to cover their energetic requirements. The ultimate result would be the pollination enhancement.

Future Prospects

Though the physiological studies on pollen behaviour have been worked out enough, but the biochemical and biophysical factors relating to its growth and development are not clearly understood. Similarly the specific needs of different tree pollens for their successful germination for different tree improvement, programmes *in vitro* conditions are not known. The pollen-pistil interaction in different environmental conditions is directly related to fertilization and is highly variable. Therefore, more studies are required to work out these phenomena. The inhibition mechanisms in self-in compatibility is a great barrier in fertilization process and the details, need to be investigated. The carbohydrates, fats, proteins and inorganic compounds present in different amounts in various grains will help in understanding the real chemical composition and the function of the grains accurately. The knowledge of Xenia and Metaxenia, pollen irradiation cytology and crossability patterns will be of immense help in different hybridization programmes of the trees. Combination of favourable traits will help very much the forestry in producing improved varieties. The technique of gene transfer through irradiated pollen in association with techniques of *in vitro* fertilization, seems to have great potential in transferring genes between distantly related species. It is, therefore, worthwhile to screen large number of genotypes in interspecific crosses as there is considerable genetic variation in their crossability.

REFERENCES

Baker, H.G. and Baker, I. 1979. Starch in angiosperm pollen grains and its evolutionary significance. Amer. J. Bot. 66: 591-600.

Brewbaker, J.L. 1957. Pollen cytology and self-incompatibility systems in plants. J. Hered. 48 : 271-277.

Brewbaker, J.L. and Emery, G.C. 1962. Pollen radiobotany. Radiat. Bot. 1 : 101-154.

Dugger, W.M. 1973. Functional aspects of boric in plants. In : E.L. Kothny (ed.), Trace elements in the environment. Advances in Chemistry Series 123, American Chemical Society, Washington DC, 112-129.

Edwardson, J.R. 1970. Cytoplasmic male sterility. Bot. Rev. 36 : 341-420.

Erdtman, G. 1954. An Introduction to Pollen Analysis. Chronica Botanica. Waltham, Mass, U.S.A.

Gaur, R.D. and Sharma, C.M. 1985. Morphology, viability and effect of growth regulators on the germination and tube growth of *Prunus cerasoides* D. Don. pollen. J. Tree Sci., 4(1) : 30-33.

Johnson, L.P.V. 1945. Reduced vigour, chlorophyll deficiency, and other effects of self-fertilization in *Pinus*. Canad. J. Res. 23C : 145-149.

Khurana, D.K. and Khosla, P.K. 1969. Pollen viability and germination studies in some hardwood species. Ind. J. For., 2(2) : 191-192.

Langner, W. 1959. Selbstfertilitat under Inzucht bei *Picea* Omorika (Pancic) Purkyne. Silvae Genetica, 8 : 84-93.

Larson, C.S. 1956. Genetics in Silviculture. Fairlawn, N.J., Essential books. p. 224.

Laser, K.D. and Lersten, N.R. 1972. Anatomy and cytology of microsporogenesis in cytoplasmic male sterile angiosperms. Bot. Rev. 38 : 425-454.

Latzko, E. 1964. Genetische Stoffwechselregulation und inhre Beeinflussung durch mineralische Nahrelemente' Bayer. Landw. Jb., 41 : 259-276.

Linskens, H.F. (ed.), 1974. Fertilization in higher plants. North-Holland Publ., Amsterdam, The Netherlands.

Lewis, D. 1949. Incompatibility in flowering plants. Biol. Rev., 24 : 472-496.

Mc Kay, J.W. 1942. Self Sterility in Chinese Chestnut (*Cestanea mollissima*). Proc. Amr Soc. hort. Sci., 41 : 156-160.

Nienstaedt, Hans and Kriebel, H.B. 1955. Controlled pollination of eastern hemlock. For. Sci., 1 : 115-120.

Orr-Ewing, A.L. 1957. A cytological study of the effects of self-pollination on *Pseudotsuga menziesii* (Mirb.) Franco, Silvae Genetica, 6 : 179-185.

Petanidou, T. and Vokou, D. 1990. Pollination and pollen energetics in Mediterrenean Ecosystem. Amr. J. Bot. 77 (8) : 986-992.

Pryor, L.D. 1957. Selecting and breeding for cold resistance in *Eucalyptus*. Silvae Genetica, 6 : 98-109.

Reimann-Phillipp, R. 1965. The application of incompatibility in plant breeding. In : S.J. Greets (ed.), Genetics today Vol. III. Proc. XI Intl. Cong. Genet., The Hague, 656-663.

Sharma, C.M. 1984. Flowering and fruiting behaviour, *In vitro* germination and viability of pollen grains of some commercial Peach cultivars. Ind. Jour. Hort. 37(4) : 225-229.

Sharma, C.M. and Gaur, R.D. 1984. Studies on morphology, germination and viability of Pomegranate (*Pinica granatum* L.) pollen. Jour. of Palynol., 20(2) : 87-92.

Swaminathan, M.S. 1964. A comparison of mutation induction in diploids and polyploids. Proc. F.A.O. and I.A.E.A. Symposium on the use of induced mutation in plant breeding, Rome.

Chapter 15

Influence of Casuarina equisetifolia's Age on the Soil Characteristics raised in Uttara Kannada District, Karnataka

*B.A. Vadiraj**

Introduction

The forest in Uttara Kannada district constitute the most important essential part of the ecosystem for sustainable development. However, over the years the forest areas originally sustaining ever green, semi ever green and moist deciduous species got depleted due to rapid increasing population of both human and cattles. The biomass removals being much higher than the increment. These compel us to go for man made forest.

Casuarina species are the members of the family Casuarinaceae which includes many species of ever green trees and shrubs with wide natural distribution (Flores, 1980; Lakany, 1983). It is a good potential tree for afforestation in regions with poor forest resources. It is having an ability to tolerate inhospitable environment like salt and drought due to its rapid foliage and root growth. Because of this it can capture their sites efficiently and to play a significant ecological pioneering role (Gauthier, *et al* 1981; Clemens *et al*, 1983; Lankany, 1983). In Karnataka *Casuarina equisetifolia* is being extensively planted in west coast on sandy soils and other areas in red laterite and lathitic soils. Much research informations are available on its propagation, usefulness under different agroclimatic zones, however informations regarding its influence on soil characteristics are scanty in Karnataka. So the present study aims

(i) to determine the physico-chemical properties of soil under different age of plantations

**Indian Cardamom Research Institute (Spices Board), Myladumpara, Idukki, Kerala 685 553.*

(ii) to assess the biological status of soil under different ages of plantations

Materials and methods

Location of the sites : The experimental sites were at Uttara Kannada district which fall under the Western Ghat region of Karnataka. The annual temperature is around 28-30°C with a relative humidity of 55-60 per cent. The annual rainfall is around 1000 to 1500 mm. The type of the soil is red, lateritic with low in fertility and degraded. The taxonomic classification of the soil is *typic Kanhaplustaifs.*

Collection of the soil samples : The soil samples were collected from 0-25 cm depth as per the standard procedure. (Prit Chett, 1979), subsequently after the first monsoon shower. Then these samples were brought to laboratory and analysed for various parameters by adopting standard methods (Jackson, 1973).

Results and discussions

The results of physico-chemical and biological properties of soils under different ages are furnished in Figs. and Tables 15.1, 15.2 and 15.3 respectively.

Table 15.1 Physical characters of the soil under* casuarina equisetifolia *plantations raised in Western Ghats degraded laterite soil of Uttara Kannada (Karnataka).

Sl. No.	*Physical Characters*	*Age of the plantations*				
		One year	*Two years*	*Three years*	*Four years*	*Five years*
1.	Organic matter (per cent)	0.52	0.58	0.67	0.77	0.82
2.	Bulk density (gm/c.c)	1.11	1.11	1.18	1.20	1.20
3.	Particle density (gm/c.c)	2.40	2.41	2.38	2.41	2.41

4.	Pore space (per cent)	54.00	53.90	50.00	50.00	50.00
5.	Moisture content (per cent)	11.14	11.14	13.28	13.55	16.80
6.	Maximum water holding capacity (per cent)	51.40	55.40	60.11	62.29	69.01
7.	Volume of expansion (c.c/10 c.c soil)	2.24	2.41	2.88	2.95	2.93

Table 15.2 : Chemical characters of the soil under* casuarina equisetifolia *plantations raised in Western Ghats degraded laterite soil of Uttara Kannada (Karnataka)

		Age of the plantations				
Sl. No.	Chemical characters	One year	Two years	Three years	Four years	Five years
1.	pH (1:2:5, soil: water)	5.57	5.59	6.37	6.53	6.79
2.	Electrical conductivity (mm hrs./cm^2)	0.09	0.10	0.13	0.17	0.22
3.	Organic carbon (per cent)	0.30	0.34	0.40	0.45	0.48
4.	Available phosphorus (kg/ha)	4.37	10.63	11.63	16.96	17.70
5.	Available potassium (kg/ha)	235.20	246.40	251.60	262.67	273.20

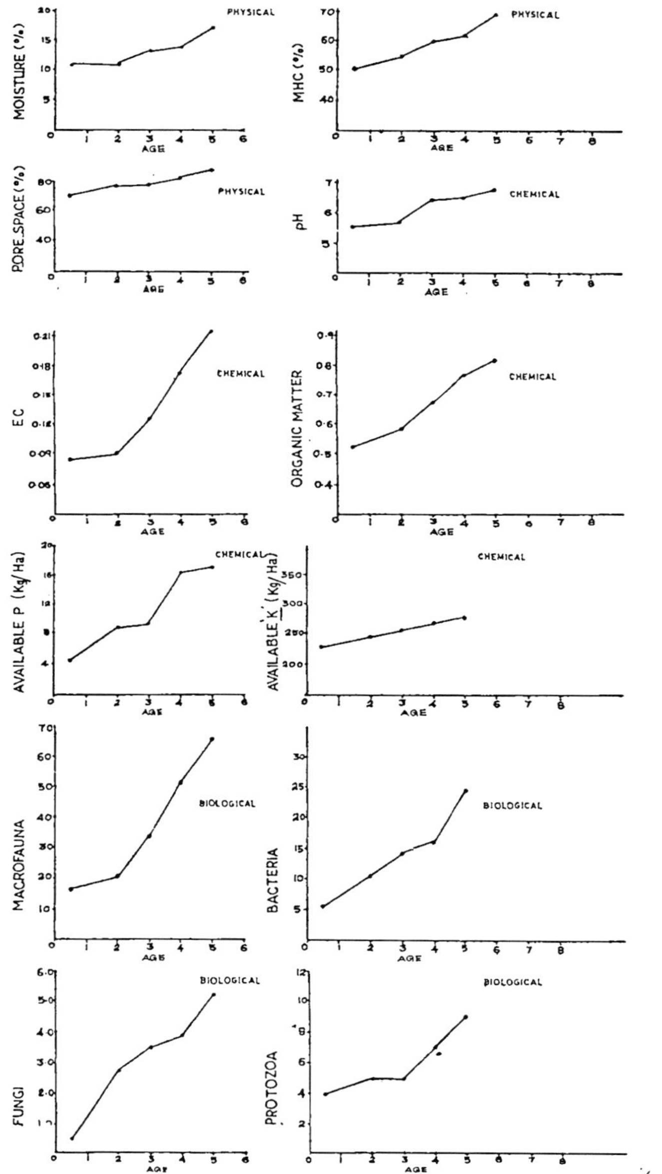

Fig. 15.1. Physico-Chemical and Biological Status of soil under Casuarina Equisetifolia Plantations

Table 15.3 : Biological status of soil under* casuarina equisetifolia *plantations raised in Western Ghats degraded laterite soil of Uttara Kannada (Karnataka)

Sl. No.	Biological status (per 1000 c.c. soil)	Age of the plantations				
		One year	Two years	Three years	Four years	Five years
1.	Macro faunal density	16.00	21.00	34.00	52.00	66.00
2.	Bacterial population x 106	5.09	10.98	14.34	16.18	24.67
3.	Fungal population x 106	0.40	2.86	3.47	3.87	5.17
4.	Protozoal species diversity	4.00	5.00	5.00	7.00	9.00

Physical properties : The physical properties studied were organic matter, bulk density, particle density, pore space, moisture content, maximum water holding capacity and volume of expansion. The values ranged from 0.52 to 0.82 for organic matter, 1.11 to 1.20 for bulk density, 2.38 to 2.41 for particle density, 50 to 54 per cent for pore space, 11.14 to 16.80 per cent for moisture content, 51.40 to 69.01 per cent for maximum water holding capacity and 2.24 to 2.55 for volume of expansion from one to five years old plantations. From these values it is quite interesting to note that important properties like O.M, B.D., moisture content and maximum water holding capacity and volume of expansion increased with increasing in age indicating the efficient accumulation and subsequent degradation resulting in effective improvement in these properties. Gauthier *et. al.* 1981; Clemens, *et. al.* 1983; Lankany, 1983 reported that *Casuarina* has the ability to tolerate inhospitable environments and has the greater potential for rapid foliage and rooting. These might contributed for improving these physical properties.

Chemical properties : The chemical properties studied were pH, electrical conductivity (EC), Organic carbon (OC), available phospherous and available potassium. The values ranged from 5.57 to 6.79, for pH, 0.09 to 0.22 for E.C., 0.30 to 0.48 for O.C. and 4.37 to 17.70 for available phosperous, 235.20 to 273.20 for available potassium from one to five years plantations. From the results obtained it is significant to note that all these properties increased with increase in age of the plantation. There are report about its ability to capture their sites efficiently by recovery of nutrients by its roots from deeper layers of the soils and also its association with beneficial micro-organisms like *Frankia* an actinomycetes and *Mycorrhiza* a fungus. These might have contributed for increase in the nutrient build up with advance in age especially phosperous and potassium.

Biological properties : The biological properties studied were macrofaunal density, bacterial population, fungal population and protozoal species diversity. The values ranged from 16 to 66 for macro faunal density, 5.09 to 24.67 for bacterial population 0.40 to 5.17 for fungal population and 4 to 9 for protozoal species diversity with the advance in the age of the plantations. These increased population of macrofauna, bacteria, fungus and protozoan species, indicates that *casuarina* provides favourable environment for these organisms.

Conclusions

Importance of afforestation to combat the ecological problems and also to meet the ever growing demand for fuel and other thing needs any emphasis. However the difficulty arises from the relegation of large tracts of infertile soils to forest plantations besides the greater demands made on soil nutrient reserves by the fast growing man made forests. However, there is an urgent need to identify and correcting the deficient sites besides increasing the biomass. From the present study it is significant to note that planting of casuarina equisetifolia in the degraded red and laterite soils not only fulfils the human needs but also improves the soil physico-chemical and biological properties. Further increase in the available phosphorus and potassium over the years indicates that these species not going to deplete the nutrients from the soils. These needs further investigations. If this is so it can be even effectively exploited for our agroforestry programmes.

Acknowledgement

The author is very grateful to Sri. A.N. Yellappa Reddy, Conservator of forest (Research and Utilization) for timely guidance and to Dr. Kubra Banu and Dr. Radha Kale, University of Agricultural Sciences, Bangalore for helping to assess the biological properties and also their valuable suggestions.

REFERENCES

Clemens, A., Campbell, L.C. and Nurisgian, S. 1983. Germination growth and mineral ion concentration of *Casuarina* species under saline conditions, *Australian. J. Bot. 31* : pp. 1-9.

E.L. Lankany, H.M. 1983. Breeding and improving *Casuarina*, a promising multipurpose tree from arid regions of Egypt. In S.J. Midgley, T.W. Turlbull, R.D. Johnson Ed. *Casuarina* ecology, management and utilization, CSIR, Australia.

Flores, E.M., 1980. Shoot Vascular system and phyllotaxis of *Casuarina* (Casuarinaceae) *Amer, J. Bot. 67*, pp. 131-140.

Gauthier, D. Diem, H.G. and Dommerques, 1981. *In vitro* nitrogen fixation by two actinomycete strains isolated from *Casuarina* nodules. *Appl. Environ. Microbiol.*, 41 : pp. 306-308.

Jackson, M.L., 1973. Ed. Soil Chemical analysis. Prentice Hall India Private Limited, New Delhi.

Chapter 16

Techniques of Contour Plantation on Denuded Hills

*Dinesh Kumar * and S.D. Bhardwaj***

Out of the 328 million ha of India's total land area, nearly 175 million ha are undergoing severe soil erosion. The menace has assumed alarming proportions in the hills which stand denuded on account of large-scale and unrestricted fellings of trees in the past. Many of these hills are watersheds and feed many river-valley projects. An analysis of the data collected by the Ministry of Agriculture, Government of India, shows that by treating 25 per cent of the catchment areas, about 50 per cent reduction in sediment production rate can be achieved (Dhuvanarayana and Sastry, 1988). It is unfortunate that vast land area is left out for any productive purpose in a country like India which has one of the smallest per capita land area availability in the world. It is thus imperative to afforest the hill slopes.

The present-day emphasis on afforestation has regrettably been confined to merely achieving the plantation targets, unmindful of the establishment success. It is a common sight in the hills to see the same areas being planted and replanted year after year with a little success. A big reason behind the fiasco is a technical one — pit planting, as is the commonest practice in the hills. Young seedlings face moisture stress and competition with weeds for light, moisture, nutrients and space in this practice. Contour planting is an effective technique for solving this problem.

Contouring is a practice in which furrow, ridge, ditch, terrace, etc. are constructed on the level, as far as practical, along the hill side, i.e., on the contour. This breaks the slope length, decelerates the surface runoff

**Deptt. of Silviculture and Agroforestry*
***Dr. Y.S. Parmar University of Horticulture and Forestry, Solan (H.P.)*

and consequently retards its scouring action and carrying capacity. By virtue of the considerable increase in water infiltration rate, worked contours are more successful in sowing and planting than the pitting technique. This technique is recommended for use in dry, slopy areas.

Contour marking

Contour may be marked with the help of any levelling instrument. Abney's level is used very often. Alternatively, the following wooden contouring frame or any of its modifications may be used.

The contour frame in the form of letter A of English alphabet, is made of 5 to 6 cm wide sawn timber pieces of 1 to 2 cm thickness. Two such pieces are joined in the shape of A of height 1.5 m with legs about 2 m apart. To keep the legs in position, they are struted with a horizontal piece on which the middle point is marked. A plumb-bob is suspended from a nail fixed in the apex (Fig. 16.1). When the thread of the plumb-bob passes through the middle-point mark of the horizontal piece, the two points on which the contouring frame stands, are on the same contour.

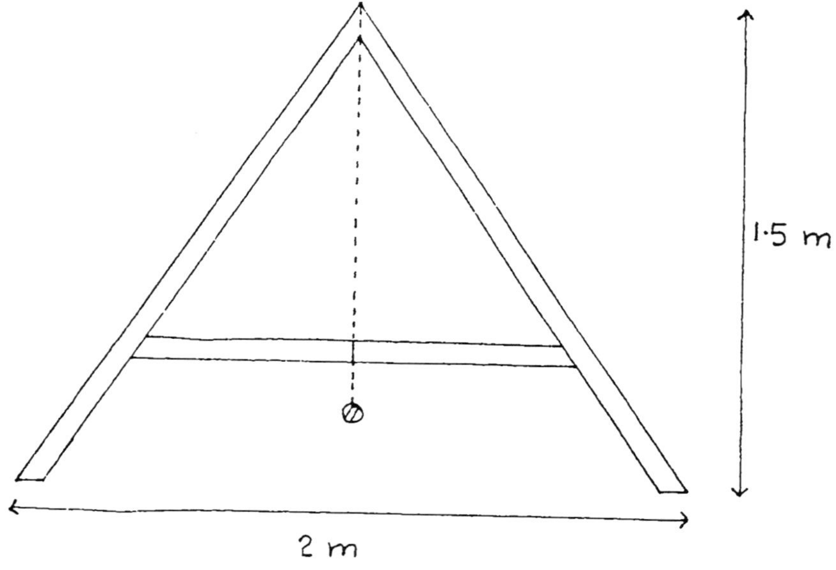

Fig. 16.1 A contouring frame with a plumb-bob.

Marking process is started from one side of the top of the slope. Having selected the starting points, a peg is fixed and one leg of the frame is placed there. The other leg of the frame is rotated slowly till the thread of the plumb-bob of he contouring frame passes through middle of the horizontal piece. A peg is driven near the base of the second leg. Now keeping the second leg at the same place, the frame is rotated to bring the first leg forward seeking another point on the same contour. When it is located, a peg is fixed there. The work proceeds this way. Soil working is done after marking a straight line between pegs (Khanna, 1984).

Soil preparation

Soil can be worked either in the form of (1) contour trenches or (2) contour terraces.

Contour trenches

Contour trenching implies excavating trenches along the contour. The trench may be continuous or interrupted. The latter type is considered better because the slightest defect in marking the contour may convert the former type of trench in a nullah to accelerate erosion. Continuous trenches are more useful in low-rainfall areas and the interrupted ones in high-rainfall areas. Interrupted trenches may be staggered or in line. The trench dimensions differ according to the soil, terrain and rainfall characteristics. As a matter of convenience, individual trench may be 3 to 6 m long and 30 to 60 cm deep. The distance between adjacent trenches on the same contour is also variable. The base of the trench should be level. The distance between successive rows of contour trenches may be 2 to 5 m depending upon the gradient.

The trenches may be either filled in a ridge-ditch pattern with a continuous ridge of earth or they may be left empty for holding water and soil, and a ridge is made about 15 cm away from the trench on the downhill side. The different kinds of trenches suitable for adoption in different parts of the country have been described by Ghosh (1977).

In the hilly tract of Chhota Nagpur and Santhal Praganas of Bihar, about 4.6 m long contour trenches of 45 cm x 45 cm size are dug with interruption of 90 cm in between two trenches on the same contour. The distance along the slope is also kept about 4.6 m. Where the slope exceeds 10 per cent, the size of the trench is increased to 60 cm x 60 cm. The excavated earth is piled up on the lower side of the trench after hoeing the surface to a depth of 15 cm in a strip 30 cm wide, leaving on a berm of at least 23-25 cm from the edge of the trench (Luna, 1989). For raising *Acacia nilotica* plantations, interrupted contour trenches 45 cm x 45 cm in cross-section and of various lengths and spaced 3 to 5 m apart are dug and the soil allowed to weather for some time. In very dry localities, the cross-section is increased to 60 cm x 60 cm. Before sowing, the trenches are half filled and loose earth heaped on the lower side so that there is a ridge on the lower side of the trench. Sowing is done in two lines, the first on the loosened soil in the bottom of the trench and the second on the berm of the trench, at the break of the monsoon. In the years of exceptional drought, seedlings inside the trench do much better than those on the berm but the reverse is the case if a spell of continuous rainy days is experienced. On easy slopes contour bunding is combined with contour trenches to ensure maximum conservation of water. This method has proved very successful in dry areas in Uttar Pradesh and Rajasthan (Joshi, 1983).

In Maharashtra where the slope is steep to precipitous, staggered contour trenches about 3.7 m in length with a cross-section of 60 cm x 30 cm and spaced about 3.7 m trees apart are made. Where the trench site comes at the mouth of a gully, one trench even more than 3.7 m in length may be made. The horizontal distance between the trenches varies from 9 to 12 m. While refilling, the trenches are given an inward slope varying from 1 in 5 to 1 in 2 according to the local conditions. In some places, continuous contour trenches with a cross-section of 60 cm x 45 cm are dug at intervals of about 20 m. All stones and pebbles are removed and heaped on the lower side of the trenches to form a retaining wall. Soil in the trench is inwardly sloped. Sowing is done on the ridges (Joshi, 1983).

Contour terraces

Contour terracing is widely practised in hill farming but its forestry application has remained largely confined to tree planting along field boundaries, on old fallows and, more recently, to rows alongwith agricultural crops at proper spacing. All types of terraces employed in agriculture can be used for tree raising, but the diversion of the less steep slopes to farming leaves only bench terracing as the terrace technique of practical use in exclusive tree plantations (Plate 16.1). Bench terracing should be avoided where grasses are absent and soil becomes very soft after rainfall.

The length and width of a contour terrace and the distance between successive contour terraces depends upon the slope. Correct contour marking is prerequisite for making long terraces. Plants may be raised either by direct sowing or by pit planting of nursery-raised stock along the terrace at the desired spacing.

Plate 16.1 : View of a hill having continuous bench-terraces

Contour trenches and terraces should be covered with cut grass, foliage, straw or any other mulch material during dry season to effect maximum moisture conservation.

Plant response

Although many soil working techniques have been attempted in different parts of the country, data amenable to statistical analysis appear lacking. Nonetheless, the superiority of contouring over the traditional pitting practice has been ubiquitously felt. Firdaus (1944) had reported that planting in deep pits was unsuccessful on an area entirely devoid of tree growth. However, extensive contour trenching on gradoni system and planting the trenches with both conifers and hardwoods was highly successful; the plants exhibited much high percentage of survival and faster and vigorous growth than those planted in pits. In China, Chang (1960) while observing the comparative performance of young plants in holes and contour ditches noticed that the contour ditch method prevented surface runoff, provided adequate water to young plants and resulted in

80-90 per cent survival. Similarly, the general trend of *Acacia nilotica* growth performance was the best in 1.83 m long, 0.61 m wide and 0.61 deep trenches with double trench system of refilling (Kaul and Gyanchand, 1966).

Bhardwaj (1991) obtained statistically workable data on the performance of *Acacia catechu* and *Celtis australis* in the experimental area of the University at Solan in Himachal Pradesh. Contour trenches proved significantly superior to pits in respect of collar diameter, aerial biomass, root weight increment and total root length of the outplanted nursery-raised stock of the two species. The plants raised in contour terraces were taller and more vigorous than those in the pits, but the statistical significance of the superiority of the former over the latter was not demonstrated. Pits, contour terraces and contour trenches always had an ascending order of soil moisture content during the dry parts of the year. The differences were most striking during the driest period — second fortnight of October and first fortnight of November — when the soil moisture in terraces and trenches was about double than in the pits.

In a similar experiment at Solan, *Toona ciliata* seedling survival after two years of planting has been 30 to 40 per cent higher in staggered trenches than in pits having equal breadth and depth (Anonymous, 1989).

Other advantages

Besides conserving soil and moisture, contouring may provide other benefits as well. The infiltrated water softens the underlying rocks and thus accelerates the weathering process of the rocks, thereby increasing the soil depth. The plant root system develops better as the roots can penetrate softened rocks with a greater ease. Experience has shown that weeds do not invade the worked trenches and terraces for 1 to 2 years from the time of soil working and, therefore, mortality on account of weed competition remains minimal. The inadvertent injury to the seedlings during grass cutting is also reduced by the very presence of seedlings in conspicuous rows. The worked contours may serve as the site for the downhill-rolling heavy seeds to settle. This may facilitate increased and uniform natural regeneration of the species. Owing to the same reason, broadcast sowing is more suited to contouring than pitting.

Economic consideration

Contouring undoubtedly requires greater initial investment than pitting for planting an equal number of seedlings. The cost margin is *inter alia* a function of plant spacing to be adopted and the slope to be afforested. But this should not discourage the adoption of contour planting as it may later prove more remunerative than pit planting on account of the greater plant survival, increased yield of leaf, small wood and other annual produce, and a shortened rotation age resulting from the more conducive growth conditions created by the former planting technique. In the light of *Acacia catechu* and *Celtis australis* growth data, Bhardwaj (1991) projected that the high initial cost of contouring *vis-a-vis* pitting could be more or less compensated due to the higher plant survival and the expected shortening of tree rotation in contours. This study was conducted in regularly-weeded area. On slopes having little or no weeding provision, contouring may pay even more relative dividends.

Conclusion

Tree raising along the contour, using either trenches or terraces, is an effective, time-tested means of successful raising of tree plantations in hilly and other slopy areas. The technique increases water infiltration, thereby increasing water availability to the plants, besides providing other benefits. The technique seems to be economically advantageous considering the reduced mortality and shortened rotation, achievable as a sequel to contour planting.

REFERENCES

Anonymous, (1990). Annual Research Report : 1988-89. Dept. of Silviculture and Agroforestry, Dr.Y.S. Parmar University of Horticulture and Forestry, Solan, pp. 1-30.

Bhardwaj, D.R. (1991). Studies on soil working techniques and planting methods for *Acacia catechu* and *Celtis australis*. M.Sc. Thesis. Dr. Y.S. Parmar University of Horticulture and Forestry, Solan.

Chang, Y.L. (1960). Observation on the growth of young plants and on prevailing surface runoff, with planting in holes and contour ditches. *For. Sci.* Peking No. 2, pp. 125-131.

Dhuvanarayana, V.V. and Sastry, G. (1988). Integrated watershed management in high rainfall areas — experience in the Himalayan region. In : National Seminar on Water Management : The key to Developing Agriculture (ed. J.S. Kanwar). New Delhi. Agricole Publ. Acad. pp. 523-542 ISBN 81-85005-54-0. CSCRTI, Dehradun.

Firdaus, A.A. (1944). The afforestation of Shankaracharya hills. *Ind. For.*, 70(2), pp. 83-85.

Ghosh, R.C. (1977). *Handbook on Afforestation Techniques.* Controller of Publications, Delhi.

Joshi, H.B. (1983) (ed.). *Troup's Silviculture of Indian Trees.* Vol. IV. Controller of Publications, Delhi, pp. 345.

Kaul, R.N. and Gyanchand. (1966). Response of *Acacia nilotica* ssp. *indica* seedling to progressive spacing, type and time of soil working. *Ann. of Arid Zone.*, *5*(1), pp. 25-35.

Khanna, L.S. (1984). Principles and Practice of Silviculture. Khanna Bandhu, Dehradun, pp. 371-372.

Luna, R.K. (1989). *Plantation Forestry in India.* International Book Distributors, Dehradun, pp. 379-380.

Chapter 17

Pathogen Management under Agroforestry Systems

*J.C. Kaushik and S.S. Bisla**

Introduction

A widely accepted definition of agroforestry is "A sustainable land management system which increases the yield of land, combines the production of crops including tree crops and forest plants and/or animals simultaneously or sequentially on the same unit of land and applies management practices that are compatible with the management practices of the local population (King, 1979). One of the main objectives of agroforestry system is to produce healthy industrial and domestic wood. To achieve this it is necessary to keep the trees free from pathogens. This can be achieved by following proper and suitable pathogen management practices which should be economically feasible. The pathogen situation varies with the nature of agroforestry ecosystems viz., shifting cultivation (Jhum), taungya system, tree plantings in agricultural fields and multipurpose tree production on canal side, roadside, railway line plantings. Pathogen management in agroforestry systems occupies an important position for making agroforestry a success. Though little work has been done to manage these pathogens, an effort is made to highlight the nature of pathogen problems in different agroforestry systems being practiced in India and the methods of their management.

Pathogen Management in Agroforestry Nurseries

Agroforestry nurseries provide ideal conditions for multiplication and survival of the pathogens due to availability of host throughout the year. Seedlings in agroforestry systems are attacked by a large number of pathogens. These pathogens cause heavy losses not only in seed beds and

* *Department of Agroforestry Haryana Agricultural University Hisar.*

container seedlings but also determines their success when outplanted, as the infected planting stock rarely survives under field conditions. To obtain healthy stock it is necessary that disease management becomes an integral part of the nursery management. A brief account of nursery diseases and their management is given below :

Nursery Diseases

Agroforestry nurseries are generally raised on agriculture or fellow lands. These seedlings raised in these nurseries is, therefore, greatly exposed to pathogen attacking agricultural crops though forest crop pathogens may also get introduced through seeds, seed cheff, leaf mulch and cause serious damages. Once the seed has been sown, it is likely to be attacked by a number of pathogens present either on the seed itself or in the soil or on debris or on living plant material.

Damping off

The first disease that occurs after sowing of seeds is damping off. It is widely distributed all over the world and probably the most serious of all nursery diseases. Two stages of the disease may be recognised i.e. pre-emergence damping-off where the pathogen decays the seed or kill the seedlings before their emergence from the soil and post emergence damping-off where the seedlings are killed after these appear above the ground. A large number of pathogens are responsible depending upon the host species and location of the nursery. Some of the important pathogens reported to cause damping-off in tree seedlings are the species of *Fusarium, Rhizoctonia, Cylindrocladium, Pythium, Cladosporium, Botrytis, Botryodiplodia, Aspergillus, Penicillium* etc. (Sharma *et al.*, 1985; Boyce, 1961; Browne, 1968). Under Indian conditions post-emergence damping-off is more common in tree seedlings (Singh, 1988; Singh and Singh, 1987).

Nursery root rots

This disease is characterised by the toppling down or falling over of the seedling but not the normal type of wilting which is mainly due to the stoppage of water supply. It commonly occurs when the tissue of the stem just above the soil become decayed while the rest of stem is still fairly turgid. The most common fungi that cause root rot in nurseries are species of *Phytophthora, Helicobasidium Rosellina, Macrophomina* and *Cylindrocladium* (Gupta *et al.*, 1988). *Rosellinia aquilla, R. necatria* and

R. quercina cause root rot in mulberry and oak seedlings (Kulkarni *et al.*, 1979). Singh and Chauhan, 1984 observed charcoal rot or black root rot caused by *Macrophominia* in *Eucalyptus*, *Acacia*, *Casuarina* and *Bombax* sp. Kaushik and Gupta (1988) observed severe root rot in *Eucalyptus* due to *Cylindrocladium quinqueseptatum*. Bakshi *et al.* (1972) reported that *Ganoderma lucidum* caused 100 per cent mortality of poplar seedlings in certain agroforestry nurseries.

Web blight of eucalyptus

It is caused by *Rhizoctonia solani* particularly in 1 to 2 months old seedlings. Infective mycelium of the pathogen emerging from soil grows up on the stem and leaves. Later from these seedlings mycelial strands arise and entangles nearby healthy seedlings giving a characteristic cob-web appearance. It was first reported from Kerala State (Sharma *et al.*, 1985).

Poplar butt rot

Bakshi *et al.* (1972) have observed that the newly planted cuttings of *Populus deltoides* start rotting after emergence of leaves. Disease is more severe in soil infected with *Botryodiplodia palmrum* in comparison to un-infected soils.

Bacterial wilt

This disease was recorded in *Casuarina equistifolia* and *Tectona grandis*. The bacterium *Pseudomonas solanacearum* causes heavy losses in these crops. Initially, the small patches of brown tissues appear on leaves between the veins. This is followed by wilting of leaves and seedling death. The dark discolouration of vascular tissues confirms the disease.

Crown gall

The bacterium *Agrobacterium tumefaciens* induces uncontrolled growth of cells in a number of tree species. Hypertrophy and hyperplasia results in tumour formation. The pathogen survives in soil and enters the host plant through wounds.

Foliage and stem diseases

The foliage and stem pathogens mainly fungi and bacteria are common to both nurseries and plantations. A few of them are discussed below.

Grey mould

Though the disease is known to be serious in conifer nurseries, the seedlings of *Eucalyptus, Rhododendron* and *Ailanthus* are also affected severely. *Botrytis cineraria* is the causal agent. The perfect stage of this fungus is *Sclerotinia fuckeliana.* Seedlings show top die back symptoms with yellowing to browing of the leaves which get covered with fungal conidiophores and conidia. The pathogen is more serious because of its ability to grow and survive under diverse conditions.

The other foliar and stem pathogens that cause leaf spots, powdery mildews rusts and twig or stem blights in different agroforestry nurseries as well as in plantations are listed in Table 17.1.

Table 17.1 : Foliar diseases of trees under agroforestry systems

Disease	*Host(s)*	*Pathogen(s)*
Anthrocnose	Ash, Oka, Sycamore Shisham	*Gnomonia* sp.
Leaf gall	Azaleas, Rhododendrom	*Exobasidium-vaccinii*
Leaf blister	Poplar and Willow	*Taphrina* sp.
Powdery mildew	Sissoo	*Phyllactinia delbegias*
	Morus	*P. moriacala*
	Poplar	*Uncinula salicis*
	Eucalyptus	*Oidium eucalyptii*
	Khair	*Erysiphae acaciae*
Leaf spots	Ecalyptus	*Cylindrocladium quinqueseptatum*
		Alternaria alternata
		Phaeseptoria sp.
		Phomopsis sp.
	Siris	*Cercospora albiziae*
	Casuarina	*Phoma casuarinia*
	Mulberry	*Cercospore mori*
		C. maricola
	Acacia	*Calonectria theae*

Poplar	*Cercospora populina*
	Septoria populi
	Sphaceloma populi
Sissoo	*Cercospora sissoo*
	Colletogloeum sissoo
	Phyllosticta sissoo
Neem & Bakain	*Cercospora meliae*
	C. subsellis
Cassia siamea	*Alternaria alternata*

Management of nursery diseases

Nursery diseases like damping off and root rots are caused by such a large number of widely unrelated pathogens that it is not possible to give exact measure that will be applicable under all conditions. An integrated disease management approach is, therefore, suggested comprising adaptation of appropriate nursery practices, seed and soil sterilization before sowing and timely use of fungicides, bactericides and other chemicals. A brief description of these practices is given.

By adapting nursery practices

Nurseries should be located on well drained, light textured and fertile soils, for heavy and alkaline soils can cause disease severity by delaying seedling emergence, thereby giving more exposure of young sucullent seedlings to the pathogens (Peace, 1962). Generally the pathogens survival is enhanced and effectiveness of chemical control is reduced in heavy soils. Such soils should be avoided or mixed with sand and organic matter (Garrett, 1970). Bakshi (1976) reported that bacterial root rot of teak and casuarina seedlings can be managed effectively when seedlings are raised on well drained light textured soils. Seedling blight of eucalyptus caused by *Cylindrocladium* sp., can be effectively managed by growing in a mixture of soil, sand and organic matter in equal proportion (Kaushik, 1988). Excessive soil moisture should be avoided and the seed bed should be maintained well aerated. Excessive soil moisture weakness the seedling which become more prone to pathogens (Griffin, 1963; Kaushik, *et al.*, 1989).

Deep and dense sowing increase the incidence of a damping-off (Gibson, 1956). In most nurseries shallow sowing to a depth of about 1/ 1/2 times of the diameter of seed and removal of covering soil from the crowns of seedlings give control of this disease (Peace, 1961 and Sehgal, 1983). A medium density of about 200 seedlings per 30 cm^2 of area of eucalyptus has been recommended. (Sharma and Mohanan 1987). Both constant monoculture and raising nursery on same piece of land for long periods also leads to flare up of diseases. If possible, nursery site should be changed regularly to avoid severe losses. Seed sowing of poplar in finely divided sphagnum moss prevented damping off disease.

Solarization

Energy from sunlight trapped in the soil elevated temperature to a level which is high enough to inactivate soil borne pathogens. The energy of soil radiation is trapped when clear polyethylene sheet is spread over the soil, because the reradiation energy of longer wave length does not pass through these sheets. The Temperature by soil solarization treatments ranges between 25 to 52°C which is high enough to reduce the potential of soil borne pathogens. (Katan, 1981). The nursery plots covered with polyethylene sheets for soil solar heating up to 55 days had 64 to 75 per cent less incidence of damping off and root rot pathogens i.e. *Pythium,* Fusarium, Rhizactonia (Hildebrand, 1985), indicating the vast potential of this source of energy in agroforestry nurseries for disease management.

Biological control

Management of some nursery pathogens, through biological means have been proved very effective. Nesme *et al.*, (1987) reported that crown bacterium of poplar was successfully controlled by *Ag.'obacterium radiocter* strain K-84 which produced a bacteriocin "agrocin 84", which was very effective against this pathogen. The other antagonists like species of *Trichoderma, Bacillus subtilis* and *Entrobacter* sp. were found effective against a number of soil borne pathogens (Wright, 1956; Kaushik, 1988; Mukhopadhaya *et al.*, 1988). Mycorrhiza roots are well known, both in agricultural and forest crops, that provide resistance to infection by various root rot pathogens (Zakes, 1964; Lakhanpal, 1988).

Chemical control

Chemical control is generally expensive and practicable only in nurseries and plantation crops where incidence of disease is high. They are applied in several ways as seed treatment, soil treatment and foliar sprays or dusts.

Seed treatment

Some seed and soil borne fungi attack the developing seedlings just after its germination and cause considerable losses. Seed treatment provides protection against both seed borne and soil borne pathogens. Seed disinfection is certainly one of the most promising means of disease control in agroforestry nurseries, especially if its influence could be extended to last until the seedlings had grown beyond serious damage. Damping off of eucalyptus seedling caused by *Cylindrocladium clavatum* was controlled by seed treatment @ 0.2% with argoll-3 or captaf or thiram (Rattan and Dhanda, 1988). Sharma and Mohanan (1981) reported control of damping off agent with Bavistin, Dithanem-45, Ferbam and Zineb when applied as seed treatment in eucalyptus.

Soil treatment

A number of fumigants like methyl-bromide, ethylene dibromide, formaldehyde and chloropicrin have been successfully used for controlling a number of soil borne pathogeny through soil sterilization before sowing. These chemicals are highly phytotoxicant and should be applied in nursery beds atleast 20 days before sowing by which time toxic effects of these chemicals are removed through volatilization. Diseases of agroforestry nurseries can also be managed by soil application of fungicides at the time of sowing or after sowing. However, if the fungicide is applied at the time of disease appearance, such treatment will be less effective. Out of many chemicals tried earlier, Thiram, Zineb, Benlate, Bavistin, Dithane-m-45, Emisan, Captaf and Difolatan were found effective in controlling damping-off, seedling blight, and root rots in agroforestry nurseries (Sharma and Mohanan, 1981; Bakshi, 1976; Bolland, *et al.*, 1986). Kaushik, 1989 reported effective control of seedling blight of *Eucalyptus tereticornis* caused by *Cylindrocladium quinqueseptatum* by Benlate, Bavistin, Captaf and Difolatan through seed treatment. Charcoal root rot and stem rot of *Acacia* and *Eucalyptus* caused by *Macrophomina phaseolina* was effectively controlled by Bavistin and Dithane M-45 through soil application. Root rot of mulberry seedlings caused by *F. solani* can be managed by soil application of methyl-isothiocyanate.

Foliar applications

Chemical sprays with Bordeaux mixture (4:4:50) have been recommended against a number of foliar diseases of forest and fruit tree species as early as in 1950. Bertus (1976) obtained varying degrees of control of foliar and stem blights by using Benomyl, Bavistin, Thiabendazole and Thiophenate methyl. Bolland *et al.*, 1986 found complete control of seedling and shoot blight of eucalypts when sprayed with mancozeb and copper-oxychloride. Treatment of Cryptomeria seedlings with growth regulator, maleic hydrazide increases the resistance to grey mould attack caused *Botrytis cinerea.* Jamalludin *et al.* (1985) reported complete control of *Phycoseptoria eucalypti* pathogen causing leaf spot of eucalyptus when sprayed with mancozeb in the Ist week of July and last week of September. Powdery mildew diseases can be effectively controlled by foliar application of elemental sulphur, sulfex and calixin fungicides (Nene and Thapliyal, 1979).

Plantation diseases and their management

In social and agroforestry programmes disease problems are likely to increase due to monoculture and frequent intervention of man. Mary earlier workers have recorded a number of pathogens on forest trees (Browne, 1961; Bakshi *et al.*, 1972; Sharma *et al.*, 1985) under Indian conditions. However, the information with regard to their incidence and extent of damage under agroforestry systems is lacking. The pathogens recorded in various agroforestry systems in different parts of country are discussed hereunder as root pathogens, stem pathogens and other foliar pathogens.

Root pathogens and their management

Ganoderma root rot

It is caused by *Ganoderma lucidum.* This is one of the most serious diseases of agroforestry trees. It is widely distributed and has wide host range. It causes serious losses to Khair, Sissoo, eucalyptus, albizia, poplar, etc. Where plantations are raised after clear felling in mixed deciduous forests. The pathogen exists in an endemic state in social forestry plantations along road side, railway lines and panchayat lands but develops a high inoculum in stumps and residual roots when these plantations are harvested for raising new plantations. The pathogen gets transferred from infected root mass of the previous crop to new crops when these are raised on such sites. Disease spread also take place through contact between roots of diseased and healthy trees resulting in high mortality. (Bakshi, 1976). Sissoo is resistant to *G. lucidum* during first 8-10 years of plantation age. The species is however subsequently attacked and killed by the disease. Severity of attack differs in accordance with soil texture. The soil having high clay content resist for the spread of pathogen whereas in light textured soils disease spread is rapid and affected trees are killed outrightly (Bakshi, 1972). There are some species viz, Semul, ailanthus, teak and kanju which are found resistant to *Ganoderma lucidum*. Thus planting of these tree species in between the purelines of susceptible tree species checks the spread of disease. Singh and Khan (1979) observed Ganoderma root rot in poplar in Dehradun areas in very severe form.

Fusarium wilt of Shisham

Sissoo suffers heavily to this disease caused by *Fusarium Solani F. sp. dalbergiae*, in different agroforestry systems. The disease incidence is higher when the species is raised on stiff and clayey soils with poor drainage, as on such sites finer roots get killed due to asphyxiation and are subsequently colonized by the wilt pathogen which then progress to conducting tissues and cause wilting. Sissoo under taungya plantations suffer from the same pathogen especially when raised on sites with high water table. Sandy and sandy loam soils with good irrigation facilities should be selected for sissoo plantations to get rid-off this pathogen (Singh, 1980).

Macrophomina root rot

Tree species like mulberry, poplar, acacias, albizia, fig, etc. suffers seriously from *Macrophomina phaseoli* under different agroforestry systems. The trees raised on degraded and denuded areas, and other such refractive sites like village fallows, panchayat lands etc. usually grow under stress conditions that predispose then to this disease. Subabool which is extensively planted in Maharashtra on such sites, suffers from two major diseases namely, ganoderma root rot caused by *Ganoderma lucidus* and gummosis caused by *Fusarium solani.*

Management of root rot pathogens

Root diseases can be minimized by maintaining trees vigour and avoiding wounds in the root system or near tree base. Once a tree is diseased it is difficult to control root disease. In severe cases where most of the roots are dead and the trunk is almost girdled removal of such trees is recommended. Where only a few roots are infected, thereby may delay the disease progression almost indefinitely. Since infected trees are often suffering from some previous stress, the vigour of the tree must be restored. Regular watering and fertilization are recommended as well as prunning of the crown to balance the root loss. If infected bark occurs on buttress, roots and trunk, it should be excised to healthy wood. Replanting in the same area following death from root disease should be preceded by removal of as much of the dead stump and roots as possible and the soil should be replaced or fumigated. Soil fumigation or trenching around the infected tree should also be done to prevent rest contact between diseased trees and the adjacent susceptible trees. Deep soil drenching with fungicides like, thiram, captafol, PCNB and benomyl have also been found effective against certain root pathogens (Gupta, 1988 and Bakshi, 1976).

Stem pathogen and their management

Eucalyptus spp. and *Populus* spp. are commonly planted on field boundaries and agro-silvi and agri-hort-silvi systems as single row plantations. Such trees suffer from tractor blade/plough damage on basal portions of stem as farmers tend to plough the land close to tree lines. *Botryodiploidia theobromae* has been found to colonized such basal wounds and cause cankers. The infection may also progress to roots and kill the affected plants (Singh *et al.*, 1983). Bakshi *et al.* (1972) observed bark burst in poplar especially when grown under water logging conditions or in areas having high water table. The fungus *Botryodiplodia palmarum* establishes on the resultant wounds and kills the cambium leading to canker development. The infected plants may die if the cankers girdle the stem.

Corticium salmonicolor B. & Br. the causal agent of pink disease is widely distributed in the tropics and sub-temperate regions and is probably indigenous in the regions where it occurs. The fungus possesses a wide host range among woody plants. In India, it is recorded on *Albizia falcataria, Azadirachta indica, Butea monosperma, Cassia siamea, Casuarina equistifolia, Eucalyptus* spp. *Hevea brasiliensis, Populus deltoides, Tectona grandis, Salix daphnoides, Tamarindus indica* etc. (Bakshi *et al.*, 1972). This pathogen is favoured by high rainfall and warm tropical conditions. Nectria canker is recorded on *Morus alba, Populus ciliata, Melia azadirach* etc. in plantations in northern India (Bakshi, 1976). The cankers appear as slightly sunken bark fissures upto 40 cm. long located mostly at or near the ground level, accompanied by slight flow of reddish sap. Another pathogen *Ceratocystis fimbriata* causes stem cankers on poplar and Neem (Kaushik and Toky, 1990). *Leucaena leucocephala* (Subabool) is also commonly planted on field boundaries in some states. Subabool suffers from gummosis and canker disease caused by *Fusarium semitectum*. The affected plants remain thin, stunted and become crooked. *F. semitectum* is a seed borne pathogen that attacks a number of host plants like *Murraya koenigii* and *Cassia tora.*

Management of canker pathogens

As it is not possible to manage stem diseases by a single measure, a package of practices has been developed to manage them. These practices can be grouped under cultural and chemical methods.

Cultural

Stem diseases can be minimized by avoiding all un-necessary wounds and by employing sound techniques for wound treatment when prunning, cabling or any other form of essential wounding is performed on trees. Cutting tools should be surface sterilized after each cut with a 70 per cent solution of alcohol. The infected tissues must be removed from healthy portions as a remedial measure. Cankers on small to medium branches should be removed by prunning. However, cankers on the trunk and large branches that can not be cut without destroying the value of a tree can be removed only by surgical excision of infected bark. (Tattar, 1989). Poor vigour of host increases susceptibility of tree to many stem diseases. In general, trees in poor vigour can not heal wounds and prevent the invasion of canker fungi like trees that have abundant moisture and balanced soil nutrients. Therefore, therapeutic treatments (vide infra), regular watering and application of balanced fertilizers are recommended to prevent new infections.

Chemical

In the control of cankers, different fungicidal paints have been tested and found effective in reducing existing wounds. The cankered portion is first sacrificed upto the woody part with a sharp edged knife and the would thus created is washed with either formaldehyde or spirit. After that the fungicidal paint is applied with a spatula covering the whole wound. Various paints including theophanate methyl, bayleton, chaubatia paint, bordeaux paste and santar paste have been found effective against these diseases (Gupta *et al.*, 1988).

Foliage pathogens and their management

A wide variety of pathogens affect the leaves of agroforestry trees. Foliage diseases are of great concern because of their adverse effects on photosynthetic activity of the plant. The effects of reduced photosynthetic activity lowers plant vigour which may make the plants susceptible to other pathogens. Infection by foliage fungi result in discrete localized necrotic areas or total necrosis and shriveling of leaves. These are grouped as below (Tatter, 1978).

Leaf spot	:	Dead area on the leaf that is well defined from healthy tissues.
Leaf blotch	:	Dead area on the leaf that often diffuses into the healthy tissues.
Anthracnose	:	Irregular deed tissue on leaf margin and/or along veins, often moving to the shoots and small twigs. Sometimes whole leaf is engulfed.
Powdery mildew	:	Superficial growth of white to grey white fungal material on leaves and shoots.
Shoot hole	:	Loss of dead area inside the shoots that results in a series of holes in the leaf.

Some of the foliar diseases, their hosts and pathogens are listed in the Table 17.1.

Management of leaf diseases

As the leaf diseases are seldom fatal to deciduous trees, the control measures for leaf diseases are usually recommended only when the health of the tree is poor or the aesthetic value of the tree is lost. Although a leaf disease might not threaten the health of a agroforestry tree, it could reduce growth of trees.

The leaf disease can be controlled in two principal ways : (1) Protection by sanitation: It has often been recommended that infected leaves be gathered and burned in the fall to remove any potential sources of inoculum for the tree. However, many leaf pathogens over winter on the tree, on dead twigs, on the bark and in the buds. In addition the spores that infect the leaves may also travel great distances through the wind, so that removal of local sources of inoculum may have no effect in reducing disease severity. In that case a farmer has to go for (2) Protection by chemicals: Protection against infection is done with a fungicide which sprayed at or before bud break and every 2 weeks following until the wet weather prevails.

Summary

The ecosystems in the agroforestry systems are so variable that pathogen management practices need to be location specific, other considerations are, pollution hazards, economics available expertise, etc. Therefore, pathogens management in agroforestry systems as discussed above only delineates broad principles. Damping-off and nursery root rots can be successfully managed by treating the propagation material, treating soil with fungicides and biological agents. Root rots and stem cankers are effectively controlled by keeping the tree vigorous, avoiding unwanted wounding, proper prunning, and application of fungicides. Use of resistant planting material, choice of planting site are the other means employed for the management of pathogens of agroforestry plants.

REFERENCES

Bakshi, B.K. (1976). Forest Pathology. Principles of Practices in forestry. Delhi pp. 400.

Bakshi, B.K., M.A.R. Reddy, Y.N. Puri and Singh, S. (1972). *Forest disease survey*. Final Technical Report : FRI Dehradun, pp. 117.

Bertus, A.L. 1976. *Cylindrocladium scoparium* on Australian native plants in cultivation, *Phytopath Z, 85:* pp. 15-25.

Bolland, L., Tierney, J.W., and Tierney, B.J. (1985). Studies on leaf spot and shoot blight of eucalyptus caused by *Cylindrocladium quinqueseptatum. Euro. J. Forest. Path.,* 15: pp. 385-397.

Boyce, J.S. (1961). *Forest Pathology*. IInd Ed. pp. 572, McGraw Hill Book Co. Inc., New York.

Browne, F.G. (1968). *Pests and Diseases of forest plantation trees.* Oxford, Clarendon Press, pp. 463.

Garrett, S.D. (1970). *Pathogenic root infecting fungi*: Univ. Press Cambridge, pp. 293.

Gibson, I.A.S. (1956). Sowing density and damping off in pine seedlings *E.Afr. Agric. For. J.*, 21: pp. 183-188.

Griffin, D.M. (1963). Soil moisture and the ecology of soil fungi. *Biological Reviews* 38: pp. 141-166.

Gupta, G.K. (1988). Fungicides in management of plant diseases: Importance and Problems. In: *Tree protection*, pp. 103-121. (V.K. Gupta and N.K. Sharma, eds.) ISTS, Solan.

Gupta, V.K., Sharma, R.C., Garg, R.C. and Kaushik, J.C. (1988). Forest plant diseases and their management. In: *Tree Protection* (V.K. Gupta and N.K. Sharma, eds.) ISTS Publications, pp. 103-121.

Gupta, V.K., and Kaushik, J.C. (1988). *Cylindrocladium* sp. associated with eucalypts in Himachal Pradesh, *J. Tree Science*, 7(1), pp. 17-19.

Hildebrand, D.M. (1985). Soil solar heating for control of damping-off fungi and weeds at the colorado state forest service nursery. *Tree Planter Notes*, 36: pp. 28-34.

Jamalludin, Soni, K.K. and Dadwal, V.S. (1985). Leaf blight of eucalyptus in nurseries. *Indian For.* III, pp. 1138-1140.

Katan, J. (1981). Solar heating of soil for control of soil borne pests. *Ann. Rev. Phytopathol.*, 19: pp. 211-236.

Kaushik, J.C. (1988). Studies on seedling diseases of Eucalyptus. *Ph.D. Thesis* - Dr. Y.S.P.Uni. Solan.

Kaushik, J.C. (1989). Forest nursery diseases and their management. Paper presented in Ann. Conference ISTS Held at HAU Hisar.

Kaushik, J.C. and Toky, O.P. (1990). *Ceratocystis fimbriata* - A new record on poplar from India. (communicated).

Kaushik, J.C., Gupta, V.K., and Sushil Sharma. (1990). Factors affecting seedlings blight of eucalypts caused by *C. quinqueseplatum.* (communicated).

Kaushik, J.C., Gupta, V.K., Saharan, G.S., and Andotra, P.S. (1989). Influence of soil temperature and moisture of radients on the severity of *Cylindrocladium* seedling blight and its control and management. In: *Pl. Path. Res. : Problems and Progress* (G.S. Saharan & M.P. Srivastava, eds.) pp. 60-64.

King, D.F.S. (1979). *Key note address—Some principles of agroforestry.* In: *Proc. of the Agroforestry Seminar.* P.L. Jaiswal (Ed.) pp. 17-26. Imphal, May 16-18, New Delhi - ICAR.

Kulkarni, S., Siddaramaiah, A.L. and Lingaraju, S. (1979). A new root rot disease of *Dalbergia sissoo. Current Research*, 8(1) pp. 3-4.

Lakhanpal, T.N. (1988). Dec. 21-22, 1989. Role of Ectomycorrhizae in tree health. In: *Tree Protection* (V.K. Gupta and N.K. Sharma, eds.), ISTS Publication, pp. 130-142.

Mukhopadhaya, A.N., Upadhayay, J.P. and Kaur, N.P. (1988). Biologicial control research on fruit and forest trees - an overview. in: *Tree Protection* pp. 143-155 (V.K. Gupta and N.K. Sharma, eds.) ISTS, Solan.

Nesme, X., Michel, M.F. and Digat, B. (1987). Population heterogeneity of *Agrobacterium tumefaciens* in galls of poplar from a single nursery. *Appl. Environ. Microbiol.*, 53: pp. 655-659.

Nene, Y.L. and Thapliyal, P.N. (1979). Fungicides in Plant Disease Control. pp. 507, Oxford and IBH Publishing Co., New Delhi.

Peace, T.R. (1962). Pathology of trees and shrubs with special reference to Britain. 753 pp. Clarendon Press, Oxford.

Rattan, G.S. and Dhanda, R.S. (1985). Leaf blight and seedling diseases of Eucalypts caused by *cylindrocladium* spp. in Punjab. *Annles of Biology*, 1 pp. 184-188.

Sehgal, R.S. (1983). Disease problems of Eucalyptus in India. *Indian Forester,* 109: pp. 909-916.

Sharma, J.K. and Mohanan, C. (1981). Chemical control of damping off and seedling and shoot blight of Eucalyptus in nursery. *Proceed. XVII IUFRO Congress.*

Sharma, J.K., Mohanan, C. and Florence, E.J.M. (1985). Disease survey in nursery and plantations of forest tree species grown in Kerala. *KFRI. Research Report*, pp. 36.

Singh, I. and Chauhan, J.S. (1984). Root rot diseases of cuttings of mulberry caused by Fusarium solani. *Indian J of Pl. Path.*, 2(1): pp. 81.

Singh, S. (1980). Mortality in sissoo (*Dalbergia sissoo*) in plantations Second Forestry Conference, Dehradun, January 16-19, 1980.

Singh, S. (1988). Management of diseases in agroforestry nurseries. National Agroforestry Workshop, Karnal July 21-23.

Sigh, S. and Khan, S.N. (1979). Impact of diseases on success of poplars in plantations. Proc. Seminar cum workshop on Silvi. management. Sri Nagar, pp. 139-141.

Singh, P., Khan, S.N. and Mishra, B.N. (1983). Some new and note worthy diseases of poplars in India. *Indian Forester*, 108: pp. 653-659.

Singh, P. and Singh, S. (1987). Pest and Pathogen management in agroforestry systems. *In: Agroforestry for Rural Needs* (Khosla and Khurana, eds.) pp. 153-177.

Tattar, T.A. (1978). *Diseases of shade trees.* Acad. Press, pp. 236.

Tattar, T.A. (1989). *Diseases of shade trees.* Revised Edition. Acad Press, pp. 391.

Wright, J.M. (1956). Biological control of soil borne Pythium infection by seed inoculation. *Plant and Soil*, 8: pp. 132-140.

Zak, B. (1964). Role of mycorrhizae in root disease. *Ann. Rev. Phytopathol*, 2: pp. 377-392.

Chapter 18

Insect Pest Management in Agroforestry — Biological Aspects

*P.K. Sen Sarma**

Introduction

Every living organism has its biological factors that normally inhibits its growth above the normal population growth rate. Biological control of pests and diseases is based on the principles of applied ecology. The fact that biological ecology is a branch of applied ecology is borne out by several known facts, but little understood for their value. The concept of biological control is now broadened to include any biotic factor that will exert a pressure on insects inimical to vegetation that is of benefit to human kind.

Classical concept of biological control envisages the employment of the natural enemies and diseases of a pest for purpose of bringing about economic pest control. It is, thus, not restricted to the use of insect parasites alone, either through introduction of exotic parasites into a new environment where they did not exist earlier or through mass breeding and release of established native parasites through augmentative release. Besides these, in recent years utilisation of temporal and special diversity of plant species in agro, forest and agroforest ecosystems has gained currency though the idea is not new. Species diversity frequently causes significant reduction of insect pest (Altiera and Liebman, 1986).

Another aspect, though not classical, is to manipulate the plant elements in the ecosystem in such a manner that it encourages built-up of the natural enemies of the pest, simultaneously creating ecological conditions that will supress population build-up of the pest. This is mainly

**Centre for Wastelands Development, North-Eastern Hill University, Shillong-739 014.*

done through encouraging beneficial plant species (perennial/annual) that support 10-12 polyphagous parasitic species and discouraging the plant species (Perennial/annual) that act as alternate hosts of pests.

It is also well known that mixed culture (polyculture) of plant species faces much lesser problem from insect pests than in monoculture comprising of even aged genetically similar population of crop.

Agroforestry which is another type of polyculture can be so designed as to control or reduce damage by insect pests. These are discussed.

Monoculture and polyculture - species diversity

Traditional system of agriculture is essentially a monoculture where a short duration even-aged crop is cultivated as a monoculture to the exclusion of other competing vegetation and genetic variation of the crop. Possibility of multiplication of insects feeding on it is the maximum in a pure crop without much genetic and age variation. Host searching and host locating by a pest are easy in monoculture, for the pest can locate oviposition and feeding hosts without any hindrance or difficulty. The pest population builds up rapidly due to sustained supply of quality and quantity of food material that also act as reproduction stimulator. The situation is, however, different in a mixed vegetal formation (polyculture) in which the total available food quantity is less for the pest and plant host finding for oviposition or feeding is often difficult on account of physical barriers caused by the presence of non-host plants. In addition, in a mixed vegetation where appropriate species diversity and composition has been formed, competition for food and feeding space may take place among harvivorous animal communities or between individuals of the same species. These regulatory factors definitely affect the abundance of the pest species often through natural regulation of the population of the pest. Competition among individuals of the same species results mainly from excessive egg-laying and consequent production of young larvae far in excess of the carrying capacity of the host plant. This often leads to internecine competition especially when the food supply is much less than needed, resulting in unusual mortality of the competing larvae and consequent decline of pest population (Beeson, 1941). It is heartening to mention that this basic principle which was/is followed in tropical forest pest management is now being experimented with for managing crop pests through multiple cropping system (Altieri and Liebman, 1986). It is emphasised that a variety of responses are expected in different models of polyculture and a single model, therefore, cannot fulfil all the requirements. A specific system model may be effective in one situation but may be totally ineffective in another situation. Cromartie (1981) has provided an excellent review on environmental control of insects using crop diversity.

Ecological aspects of biological control

As has been stated earlier, biological control should verily by treated as a branch of applied ecology in which interactions between plant hosts-pests-natural enemy complex assume the most important role especially in tropical eco-system on account of its rich biodiversity. Therefore, basic ecological knowledge is essentially required for an intelligent and effective utilisation of diverse modes for using carbohydrate diet. Water required by insect biocontrol agents is normally obtained from rain and/ or dew. Pollens, nectar and honey dew (produced by aphids, etc.) provide protein, free amino acids, sugars t the parasites and the predators. All these are obtained in nature from Plant Kingdom. This approach to augment natural increase in the enemy complex of the pest through environmental manipulation can be fruitfully experimented and applied in agroforestry by encouraging plants that provide pollen and nectar or that harbour honey-dew producing aphids. These are important for enhancing the abundance of natural enemies in an altered man-made ecosystem. That the flowering plants especially those belonging to Umbelliferae are important in the food economy of adults of certain hymenopterous parasites have been experimentally proved in Canada and erstwhile USSR (Lieus 1960, 1967, and Sen Sarma, 1972). Encouragement of undergrowth comprising of wild flowering plants in fruit orchards and forests provides sources of pollen and nectar. However, success of this promising method would depend on an accurate understanding and knowledge of adult feeding behaviour which, in case of indigenous parasites and predators, is sadly lacking in India.

Shelters provided by wild vegetation, hedgegrows, wind-breaks etc., enables insects to survive in certain areas. Adults parasites and predators are often used these shelters during stress conditions and as hibernation or aestivation sites. Hedgegrows or uncultivated lands adjacent to cropped areas have, therefore, a beneficial effect (Beeson, 1941).

Polyculture has been found to provide more pollen and nectar source to adult parasites and predators when compared with monoculture (Mai *et. al.* 1979). Agroforestry ecosystem seems to be ideal to generate data on this aspect.

Characteristics of agroforestry systems for biocontrol of pests

Agroforestry envisages growing perennials and annuals' seasonals in combination in the same unit of land in such a manner that productivity of land is exploited to the optimum for the benefit of the mankind. Ecologically, agroforestry is very distinct either from crop husbandry or tree husbandry. Agriculture of even aged genetically similar crops to the exclusion of any competing vegetation is not ecologically sound as it lacks biodiversites. Pests can multiply at much greater rapidity in such an ecosystem. However, in polyculture or mixed cropping which agroforestry aims, at insects may often find it difficult to locate the host-plant on account of the presence of non-host crops in good numbers. These companion crops may hinder the dispersal of pests by mechanical means (Root, 1973), may camouflage the host plant (Altieri and Lieman, 1980), may repel the pest due to unpleasant and unacceptable morphological features (Levin, 1973).

Agroforestry systems can also be a tool to manipulate physical conditions like light, temperature, relative humidity and wind condition in such a manner that these become unfavourable for the pests and be favourable to the natural enemies of pests. Shade from perennials can often interfere with the host searching. High humidity favours infection of the pests by entomogenous fungi (Jaques, 1983).

It is known that compared to monoculture, mixed cropping provides more pollen and nectare, provided the mixture is deliberately selected to serve this purpose. Availabilities of high quality nectar improves reproduction of parasites. Agroforestry serves this purpose the best. The agroforestry also provides increased ground cover through leaf-fall etc., which often promotes and accelerates predatory activities of nocturnal predators. The ground cover provides these predators ample hiding places during day (Beeson 1941, Dempster, 1969).

Some trees are known to exhibit repellency, feeding and/or oviposition deterrency. These, when incorporated in agroforestry systems, can have adverse effect on pests. On the contrary, plants/trees that produce attractants, arrestants, stimulants etc. when forms a component of mixtures in agroforestry may reduce feeding by pests whose feeding interest often get diverted to alternate host plants (Schoonhoven 1968). Altieri (1968) has shown that mixtures of various volatile chemicals released by different plants/trees are capable of creating confusing among pests in locating the host plants for oviposition and/or feeding.

Colonisation of natural enemies of pests

Successful colonisation of natural enemies (parasites and predators) in a new area and their effectiveness depend on inherent capability of the introduced enemies to adjust and survive in a new area and the biophysical conditions of the area. Macro and micro-climate of the area of introduction plays the most important part in the survival of introduced enemies. Among the climatic factors, temperature and relative humidity closely interact with each other. Some parasites and predators may flourish in an optimum temperature and relative humidity (Beirne, 1967; Debach, 1962; Ratcliffe, 1965). Agroforestry is also known to moderate both macroclimate and microclimate and this would be conducive to survival and multiplication of parasites and predators either native or introduced. The biotypes created by agroforestry improves biological characteristics of parasites and predators through improved nutrition and adequate shelter. It also improves the chances of survival and multiplication of polyphagours parasites on account of availability of secondary and tertiary host insects which can occur in biological diversity. This is necessary for maintaining a population reservoir of parasites/predators when their primary or definitive host is scarce. This mechanism has been successfully experimented in suppressing the populations of teak defoliator and teak skeletoniser, sissoo defoliator, etc., (Beeson, 1941 and Sen-Sarma, 1983).

A good example of environmental modification to encourage natural enemies of endemic insect pests has been given by Dontt and Nakata (1965) who provided overwintering refuges for *Anagrus epos*, an egg parasite of the grape leaf-hopper in California. They also provided alternate leaf hopper host of the parasite by interplanting wild black berries (*Rubus* spp.) to augment hosts for the egg parasite.

Agroforestry provides shelter and nesting sites to insectivorous birds and bats. These augment biocontrol of noxious insects when they damage the crops. Experiments carried out in several countries to build up requisite population have yielded positive results. In Indian agricultural fields, beneficial effects of these biocontrol agents are not discernible, the reason being either total absence or inadequate presence of insectivorous birds and bats.

Conclusion

The facts presented above clearly demonstrates the efficiency and value of agroforestry in suppressing harmful insects. However, intensive research on certain areas of agroforestry is needed. It is well recognised that traditional cropping patterns have evolved through the experiences of generations of farmers. In these cropping pattern, certain built-in mechanisms have been incorporated which militate against pest population build-up. These systems need to be studies, documented and experimented by means of on-farm research data and where necessary authentic scientific data are to be generated. The success depends largely on effective agroforestry designs. The areas that need attention are plant density and crop arrangement patterns, the nature of interactions among the crop components (perennials and annuals), including physical factors that may restrict pest migration, pests searching abilities to locate the principal host plant, macro-and micro-climatic factors that are conducive to maintain a reservoir of natural enemies, etc. In addition, the behavioural studies in respect of feeding and host searching for pests, parasites and predators are required. Further, qualitative and quantitative data are yet to be generated with regard to the flowering plants that may provide adequate and quality pollen and nectar to the adult parasites and predators. While experimenting with appropriate and effective mixtures for biocontrol of pests, even mixtures among the annuals may be considered. While selecting perennials and annuals for the polyculture, care must be taken to avoid alternate hosts of pests during selection. Here our knowledge about commonality of insect pests that might attack both perennials and annuals is abysmally inadequate and almost non-existing.

Extensive and intensive survey is needed in the existing agroforestry system to determine pest, parasites and predators complexes. If agroforestry is considered as an alternate scientific agriculture, generation of these data is essential through an experimental designing and developing appropriate models.

REFERENCES

Altieri, M.A. (1986). *Agroforestry: The scientific basis of alternate agriculture,.* 227 pp. Westview Press, Boulder Calorado.

Altieri, M.A. and Liebman, M.Z. (1986). Insect, weed and plant disease management in multiple cropping system. In *multiple cropping system,* (Ed. C.A. Francis), Macmillan, New York, pp. 183-218.

Beeson, C.F.C. The Ecology and Control of Forest Insects.

Cromortie, W.J. (1981). The environmental control of insect using crop diversity. In CRC Handbook of Pest Management in Agriculture: (Ed. D. Pimentel), Boca, Raton, Florida, pp. 223-250.

Dempster, J.P. (1969). Some effect on weed control on the numbers of small cabbage white butterflies (*Pieris rapae* L.) on Bruses sprouts, *J. Appl. Ecol.* 6, pp. 339-345.

Jaques, R.P. (1983). The potential of pathogens for pest control, *Agri-Ecosystem, Environ.*, 10 pp. 101-126.

Leius, K. (1960). Attractiveness of different foods and flowers to the adults of some hymenoterous parasite. *Canadian Ent.*, 92, pp. 369-376.

Leius, K. (1967). Influence of wild flowers on parasitism of tend caterpillar and codling moth, *Canadian Ent.*, 99, pp. 444-446.

Levin, D.A. (1973). The role of trichomes in plant defence, *Quart, Rev. Biol*, 48, pp. 3-12.

Mai, X. Huang, M., Wer, W., Ouyang, Z. Lin, Z., Zhu, R. and Li, Z. (1979). Control of citrus red mites by augmentation of natural population of *Amblysius* spp. in Hill side orchards, National enemies of Insects, pp. 52-56.

Root, R.B. (1973). Organisation of plant-arthropod association in simple and diverse habitats : fauna of collards (*Brassica oleracea*), Ecol, Mong; 43, pp. 94-125.

Schoonhoven, L.M. (1968). Chemosensory bases of host plant selection, *Ann. Rev. Ent;* 3, pp. 115-136.

Sen Sarma, P.K. (1972). Insect factors in the management of man made forest, *Proc. Symp. Man-made Forest in India*, Dehra Dun, IV pp. 81-85.

Chapter 19

Insect Pests Management under Agroforestry Systems of India

R.S. Bhandari and Sushil Kumar*

Introduction

The role of forest is well evidenced in favour of people's welfare. The forest cover of India which was about 22 per cent of the total land area has now been reduced to about 19 per cent (Lal, 1989). The quantum of goods and services rendered by the forest is enormous but the existing forest cover is unable to meet the growing demands of goods to 850 million people of India. It will be catastrophic from the ecological point of view, if we fail to maintain the existing forest cover. To meet the growing demand of forest products and agricultural commodities, efforts are now being made to increase the productivity of the same unit of land by the practicing of agroforestry. Thus, agroforestry system is a practice to bring perennial grow forest trees and annual/seasonal agricultural crops together on farmlands. Agroforestry system this is a form of intercropping in which one (or more) components is a woody perennial.

Insect is one of the most deleterious factors which is often responsible for the failure of forest tree plantations in an agroforestry ecosystem. In such an unstable ecosystem, insect population rapidly multiplies in the absence of natural enemy complex. Sometimes, an agricultural pest may become the pest of forest tree or vice-versa. Thus, an agro-forestry system become somewhat more complex owing to fluctuating trends in population of the insect pests of a agricultural crop or tree crop or both.

Seedlings and saplings of agroforestry trees are most vulnerable. Often whole of the nursery plants become the victim of insect scourage that may lead to the complete failure of the programme.

**Division of Forest Protection (Entomology), Forest Research Institute, Dehra Dun-248006.*

Agroforestry systems practiced in India in brief

In India many agro-forestry combinations are prevalent. Among which three combinations are most popular and practiced as shifting cultivation (Jhum), Taungya system and multipurpose agroforestry systems (Singh *et al.*, 1990), Khosla *et al.*, 1985). Different types of combinations are briefly as follows:

Combination-I

Trees like, *Acacia nilotica, A. tortilis, A. catechu, Albizia lebbek, A. albida, Prosopis cinerarea, P. juliflora, Tecomella undulata* are grown with all usual crops of the region like millets, sorghum, gram etc. in arid and semi-arid regions.

Combination-II

Trees like *Amoora* spp., *Anthocephalus chinensis, Ailanthus excelsa, Anacardium occidentale, Artocarpus chaplasa, Bombax ceiba, Dalbergia sissoo, Eucalyptus* spp., *Gmelina arborea, Leucaena leucocephala, Morus alba, Poplar* spp., *Shorea robusta, Tectona grandis* are grown along with wheat, rice, tapioca, maize, potato, groundnuts, pulses, beans, seasame, tobaco etc. in different parts of the country.

Combination-III

Important tree species of multiple use like *Grewia optiva* (Biul), *Celtis australis* (Khirk), *Robinia pseudoaccia* (exotic), *Morus alba, Bauhinia* spp. *Anogeissus latifolia, Salix* spp. and poplars are extensively grown in and around the agricultural fields in West Himalayan region.

Region wise, agroforestry practices as follows :-

Agroforestry practice in north-east region

Jhum or Shifting Cultivation in which an area is selected for raising crops by clear felling the vegetation and burning the felling refuse. It is practiced by the tribals of North-East, Orissa and some parts of Peninsular India. Crops are grown for 2-3 years and than the area is abandoned to restore fertility. Before leaving, tribals broad-cast forest tree seeds like Kashi pine. In Sikkim, large cardamom (*Amomux subalatum*), a traditional plantation crop, is grown with *Alnus nepalensis* (Utis). Bamboo groove of *Dendrocalamus hamiltoni* and *D. sikkimensis* is quite common with other tree species like *Aegle marmelos, Albizia lebbek, Bauhinia purpurea,*

B. variegata etc., as component of agricultural holdings.

Taungya type Agroforestry

It is a method of growing forestry crops in which a land area is allotted to a person, family or a group of persons to grow agricultural crops between the row of planted trees. They plant and protect forestry plantations and crops. This system is popular in West Bengal, Uttar Pradesh, Kerala and to a lesser extent in Andhra Pradesh, Karnataka, Maharashtra, Orissa and Tamil Nadu. Most popular, trees for taungyas are sal, teak, babul, khair, maharukh, semul, toon, champa etc.

Hill regions

The forest tree species discribed in Combination-III are grown in agricultural holdings on the boundaries of agricultural fields.

Alluvial plain regions

A number of tree species are grown for their economic considerations. For example sal, teak, champa, gamhar with paddy in West Bengal, Shisham, babul, subabul in Bengal, with wheat and rape seeds; poplar, eucalyptus, mulbury, arjun in Uttar Pradesh with wheat; shisham and eucalyptus in Haryana/Punjab.

Semi-arid and arid regions

Tree species like, Babul, Jhand, Rohida, Amla, Imli, are grown with agricultural crops with sorghum milletes, pigeon pea, black grams etc., in dry region of Haryana, Rajasthan and Gujarat.

Tropical regions

Some native trees like *Morus alba, Pongamia pinnata, Tamarindus indica, Acacia* spp. and exotics like *Eucalyptus* spp. and *Leucaena leucocephala* are extensively grown with sorghum, safflower, bengal gram etc., in Penninsular India. The cultivation of sorghum and *Prosopis julliflora* and *Casuarina equisetifolia, Eucalyptus* with paddy and *L. leucocephala* with Sisbania and regional crops are widely practiced in Maharashtra, Karnataka, Tamil Nadu and Andhra Pradesh respectively.

Humid and sub-humid regions

The cultivation of cash crops like pepper and cocoa and medicinal plants is done in selected areas with forestry trees like teak and silver oak in Kerala.

Insect Pest Management in Agroforestry Nurseries

There are numerous insect pests which cause tremendous damage to forest seedlings in nurseries. Major insects are termites, white-grubs, cutworms, crickets. Minor pests are several defoliators of Coleoptera, Lepidoptera, Orthoptera etc. Sometimes, nematode and rats take heavy toll of seedlings in nurseries.

Termites

They are most destructive pests of forest nurseries. They feed on the bark and roots of seedlings. The infested seedling begin to turn pale in colour and gradually die. Such infested seedlings can easily be pulled out with slight force. At the infestation site mud plaster covering made by termites on seedling or trees also indicate the presence of termites in the area. Important termite species which cause damage to nurseries are *Microcerotermes minor, Odontotermes indicus, O. microdentatus, O. distans, Microtermes obesi* (Thakur & Sen-Sarma, 1980). Termites are polyphagous and they can feed on crops as well as forest trees. They cause severe damage to *Eucalyptus* transplants up to 3 years which often result in total failure of agroforestry tree plantations. Seedlings of *Acacia* spp., *Albizia lebbek, Azadirachta indica, Casuarina equisetifolia* and various clones of poplar are also havily damaged.

White grubs (Cockchafers)

The grubs of this insect are subterranean in nature and are generally root feeders while adults are defoliators. They were recorded to take heavy toll of teak seedlings in Maharashtra. The culprit species was identified, *Holotrichia consanguenea* (Coleoptera : Scarabaeidae); (Vaishampayan & Bhandari, 1981). The white-grubs species of *H. consanguenea, H. insularis, H. serrata, H. problematica* cause damage in forest nurseries of Bihar, Gujarat, Tamil Nadu and Uttar Pradesh as well as to the agricultural crops. In Himalayan region *Granida albospara* (Coleoptera : Scarabaeidae) causes damage to conifer nurseries. *Anomala polita* damages seedlings of *Shorea robusta.* Grubs are found below 6 to 50 cm. of earth surface. Their head is reddish brown and body is of white colour. Under rest condition, body remains in curved position. Seedlings are killed by the destruction of the rootlets or removal of bark of the tap root by feeding. The life cycle is normally annual with a larval period of 8 to 10 months in tropical and subtropical conditions. In temperate conditions, life cycle may lasts for 2 years.

Cutworms

Greasy cutworm *Agrotis ipsilon* (Lepidoptera : Noctuidae) is a polyphagous pest and causes heavy mortality to the nursery stocks of several species. The larvae cut the stem of young seedlings from the collor region. Sometimes larvae penetrate in soil and feed on roots of plants. They have been reported to cause heavy damage in coniferous nurseries (Beeson, 1941). They prefer dry climate in comparison to damp places but in summer they aestivate due to high temperature. In plains, the insect completes many generations in a year but under temperate conditions, it usually complete only one generation. Pupation occur inside the soil and adults start emerging in the second half of May.

Cricket, *Brachytrypes portentosus* (Orthopeta; Gryllidae) is an important pest of nurseries when raised in sandy soil. Its nymphs and adults feed on the root system of seedlings and ultimately kill them.

Minor Pest

Besides the major pests agroforestry nurseries are sometimes attacked by several other pests of minor importance. Mention may be made of defoliating beetles of genera *Apoderus* and larvae of *Euproctis, Lymantria, Heliothis, Plecoptera* etc. Some sap feeder like *Oxyrachis tarandus*, mealy bug and thrips also damage the seedlings.

Non-Insect Pests of Agroforestry Nurseries

In addition to insect pests, nurseries are also vulnerable to the attack of nematodes which sometimes cause immense damage. Besides invertebrate pests, some vertebrates like rats, hares, squirrels, porcupine, deer etc., cause damage to seedlings in agroforestry nurseries. Among these, rats are most destructive. They have been reported to feed on the roots of babul, shisham, bakain, teak and *Cassia siamia* respectively. (Kumar & Thakur, 1989; Ahmad and Srivastava, 1991).

Management of the Pests of Agroforestry Nurseries

Following practices can be adopted to reduce pest population in agroforestry nurseries :-

1. Deep ploughing of the nursery soil should be done to exposed the white grubs and cutworm population to birds and sunlight. Ploughing should be avoided when beetles are on wings for which loose soil may provide suitable sites for oviposition.

2. Weedling should be avoided at the time of the emergence of beetles to avoid the eggs deposition in loose soil. It should be done atleast 6-7 weeks before spring sowing. For autumn sowing, weeding should be done in August.

3. Light traps can be used to trap winged termites and beetles during their emergence.

4. For chemical control of white-grubs, most widely accepted insecticides is Phorate (Thimet) 10 G which can be used as 200 g/bed of 10 x 1 M in teak and other nurseries (Vaishampayan and Bhandari, 1981). For the control of subterranean pests 30 ml aldrin (30 EC) or 50 ml Chlorpyrifos (20 EC) in 50 litres of water can be sprayed in the nursery bed of 10 x 1 M before sowing (Thakur *et al.*, 1989). Through the experimental studies, it was found that at nursery stage, transfer of eucalyptus seedlings in polythene bags filled with treated soil with chloropyrifos 20 EC (0.3%) is effective to protect seedlings from termites even after the transplantation in fields (Thakur *et al.*, 1990). In teak nurseries of Madhya Pradesh, basudin (diazinon) 10 G @ 200 gm/bed (10x1 M) followed by phorate 200 g/bed was found effective against white grubs of *H. serrata* (Meshram *et al.*, 1990). Sap-suckers can be controlled by spraying monocrotophos 36EC (0.01 - 0.02%) and defoliators by endosulfan or malathion (0.1 - 0.2%).

Management of Insect Pests of Agroforestry Tree Species

Khasipine (*Pinus kesiya*) : It is one of the important forestry crops under Jhum cultivation in North-East India. Epidemics of defoliators and shoot and cone borer have been recorded from this region. In Khasi and Jainta hills in Meghalaya during 1974 to 1976 outbreak of defoliators, *Etrusia pulchella* (Lep. Zygaenidae) and *Dichocrosis punctiferalis* (Lepidoptera : Pyralidae) (Plate - 19.1) have been noticed (Singh and Singh, 1987). The pine shoot and cone borer *Dioryctria castanea* (Lep., - Pyralidae) epidemic was observed in 1982 in Yachuli division of Arunachal Pradesh.

Plate 19.1 : a. Defoliated plantations of khasi pine by Dicocrosis punctiferalis *b. Defoliated plantation of khasi pine by* Etrusia pulchella *c. Larva of* E. pulchella *d. Pupa of* E. pulchella *e. Moth of* E. pulchella *f. Moth of* D. punctiferalis

***Management*:** In case of severe epidemic of *E. pulchella* foliar spray of malathion or carbaryl (0.25%) is recommended (Sen-Sarma, 1986). The another pest of Khasi pine needle binder *D. punctiferalis*, although primary pest of castor, has assume the potential pest status of kashi pine. It can be controlled by spraying 0.1 per cent carbaryl in water solution (Singh, 1990). To control shoot and cone borer *D. castanea*, spraying of insecticides in field, *viz.*, phosphamidon 0.03 per cent (1.4 ml/tree) or dimethoate 0.03 per cent (4 ml/tree) or monocrotophos 0.04 per cent (444 ml/tree), (Singh *et al.*, 1988) may be done.

Bamboos : Bamboos are attacked by some 34 species of insects in green standing condition. (Singh and Bhandari, 1988b). Among the defoliators, *Pyrausta coclesalis* (Lepidoptera : Pyralidae) is distributed all over India. Epidemics of this insect occur frequently. Most important of the shoot can clum borers are *Cyrtotrachilus dux* and *C. longimanus* (Col. - Curculionidae). The hispine borer, *Estigmena chinensis* (Col. - Chrysomelidae) damages the bamboo culms and some times 100 per cent culms may be found damaged and rendered useless (Plate 19.2). An aphid *Orygma bambusae* (Homoptera : Aphidae) and *Ochrophara montara* (Hemiptera - Pentatomidae) are two important sap suckers of bamboo clums and flowers respectively.

Management : Bamboo defoliator can be controlled by spraying 0.1 per cent folithion 50 EC in water, *Oregma bumbusae* by spraying phosphomidon or dimethoate 0.02 per cent in water. Attack of *Cyrtotrachilus* spp. and *E. chinensis* is heavy in the congested clumps, so avoid congestion of bamboo in a clump to avoid attack of these insect.

Alnus neplensis : The species of *Alnus* are attacked by many insects (Bhasin & Roonwal, 1954). Beetles of *Anomala flavipes, A. grandis, Chrysomela chlorina*, defoliate the trees. Larvae of cerambycid beetle *Batocera horsfieldi* and larvae of *Endoclita undulifera* (Lepidoptera; Hypialidae) bores the heartwood of living tree and young saplings respectively.

Management : Defoliators of *Alnus* can be controlled by spraying malathion or folithion 0.1 per cent in water. This spray is effective for 15 to 20 days.

Teak *(Tectona grandis)*

Pests : Teak is the common species for the taungyas of agroforestry system in the country. The major pests of teak taungyas are teak

Plate 19.2 : a. Cyrtotrachilus dux *on damaged bamboo b.* Estiqmna chinensis *on bamboo c.* Pyrausta coclesalis *larva d.* Pyrausta coclesalis *moth*

defoliator, *Hyblaea puera* (Lepidoptera; Hyblaeidea)and skeletonizers Eutectona mechaeralis (Lip., Pyralidae) on young and mature trees (Plate 19.3). Other pests include Phassus borer, *Sahyadrassus malabaricus* (Lip; Heplalidae) and canker forming insect *Dihamus cervinus* (Coleoptera; Lamiidae) in young plantations; and shoot borer *Zeuzera coffea* (Lepidoptera; Cossidae) on mature trees. Besides these pests, inflorescens and fruits of teak are also heavily destroyed by the larvae of *Pagyda salvalis*.

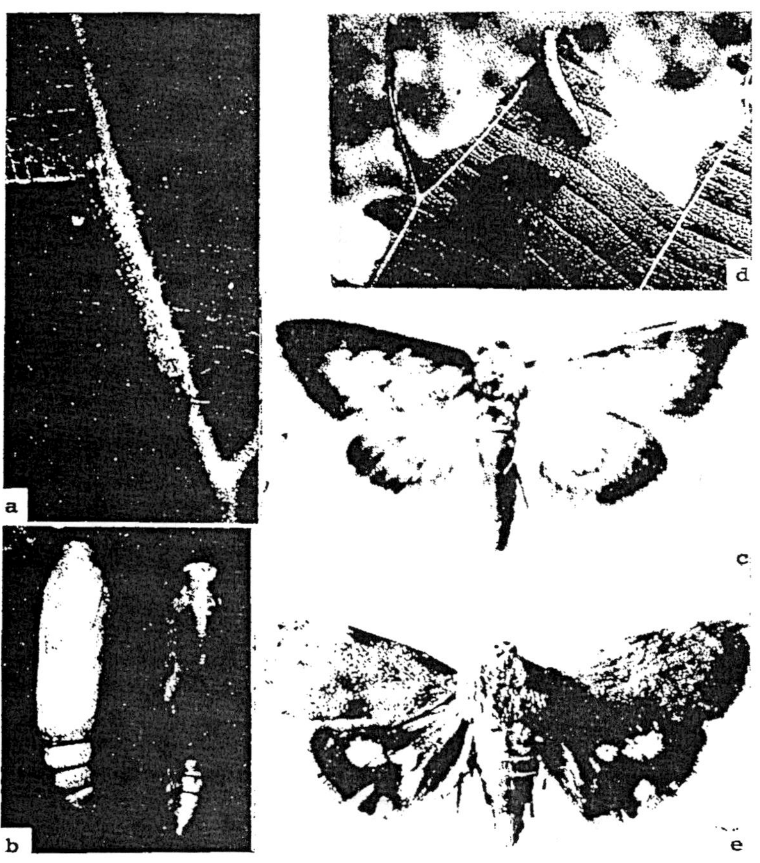

Plate 19.3 : a. Larva of Eutectona mechaeralis *b. Pupa of* Eutectona mechaeralis *c. Moth of* Eutectona mechaeralis *d. Larva of* Hyblaea puera *e. Moth of* Hyblaea puera

Management : The biological agents like larval parasite, *Cedria paradoxa* (Hymenoptera : Braconidae) and egg parasite *Trichogramma minutum* (Hymenoptera; Trichogrammatidae) have been used successfully against *E. mechaeralis* (Beeson and Chatterjee, 1939). The tests of three exotic *Trichogramma* species namely *T. evanscence, T. brasiliensis* and *T. pickel* against *E. mechaeralis* eggs in Karnataka have given encouraging results in field conditions (Patil and Thontadarya, 1983b). In nature, a nuclear polyhedrosis virue (NPV) disease causes considerable death of larvae of *H. puera* (Sudheendra Kumar, 1986). Bacteria, fungi and nematodes have been reported as natural bio-control agents of defoliators (Patil and Thontadarya, 1981, 1983a); Singh and Misra, 1987; Agarwal *et al.*, 1985; Mathur, 1959). Under genetic control methods, the planting of resistant varieties clones against the pest species is becoming popular among silviculturists. In South India, clones TNT-6 was found to be the most resistant against *H. puera.* It was suggested to have intraspecific crosses between clone TNT-6 and KLK-2 to obtain better resistant and timber yielding tree species (Ahmad, 1987). Similarly teli clone of teak was found resistant against teak skeletonizer, *Eutectona machaeralis* (Kedarnath & Singh, 1975). Studies on chemical control have also been done through laboratory bioassay. Of 20 insecticides assayed four namely, monocrotophos chlordimeform, quinalphos and anthio proved most effective against *E. mechaeralis.* (Singh and Gupta, 1978). Microbial insecticide (thuricidae) have been tested successfully against *E. mechaeralis* in laboratory (Singh & Misra, 1978), but its use in field was discouraged owing to susceptibility of silk worm in India. The borer *Cossus cadambae* have a tendency to attack the plant in patches. This behaviour is highly favourable for a pheromone based control strategy (Mathew, 1990). Microbial pathogens also cause, mortality of this pest (Mathew and Ali, 1987). The use of quinalphos at 0.125 per cent was found effective after removing frass to control phassus borer of teak (Nair, 1982). The fungus, *Beauveria bassiana* also causes mortality of larvae. The larva of *Diahamus cervinus*, a canker forming grub around the trunk can be controlled by adopting silvicultural practices. *Clerodendron infortunatum* is its collateral host. Therefore, teak should not be planted in areas previously occupied by this plant species. Damaged teak plants of 1 to 4 years should be cut at the ground level in order to obtain coppice (Beeson, 1941).

The caterpillars of *Pygeda salvalis* can be controlled by spraying 0.125 per cent endosulfan or 0.2 per cent monocrotophos in water emulsion with hormone alpha nephthyl acetic acid (NAA) 40 ppm to boost the production of seeds by protection from the insects and stimulating the growth of seeds (Neelay *et al.*, 1983).

Sal (*Shorea robusta*)

Pests : In Gorakhpur (U.P.) and North Bengal sal is grown with agricultural crops as taungyas. The major pest of sal taungyas are Lepidopterous defoliators, *Ascotis selenaria imparta* (Geomatridae), *Diacrisia obliqua* (Arctiidae) *Lymantria mathura* (Lymantridae), *Suana concolar* (Lasiocampidae) (Singh and Bhandari, 1989). Although the sal heart wood borer (*Hoplocerambyx spinicornis*) is a serious pest of sal in forest but sometimes it also attack sal in taungya systems.

Management : The defoliator *A. selenaria* can be controlled by spraying endosulfan 0.2 per cent around the trees base for killing the larvae descending from these for pupation in soil (Singh and Thapa, 1988). In nature viral disease caused by NPV take a heavy toll of larvae during rainy season. Fungal disease is also responsible for destroying the pupae of the pest (Singh and Thapa, 1988).

Shisham (*Dalbergia sissoo*)

Pests : It is widely grown around the agricultural fields in U.P., Haryana and Punjab. The tree is attacked by two major lepidopterous pests namely a semi-looper, *Plecoptera reflexa* (Noctuidae) and sissoo leaf binder *Dichomeris eridentis* (Galechidae). The most severe defoliation occurs from June-August. A loss of increment (25%) per annum is caused.

Management: To manage the pests, biological cum cultural control is practiced. To desyncronise the moth emergence with new flush, irrigation is manipulated in such a way so that flush of new leaves appears on the plant earlier than emergence of moths, and by the time emergence and hatching of eggs take place, the leaves become older and are not liked by the larvae of *P. reflexa* (Beeson, 1941). Pest infestation can also be reduced by controlled thinning of dense sowing. Introduction of *Disophrys sissoo* and *Microgaster plecopterae* parasite should be encouraged. To control *D. eridentis* and *P. reflexa* chemicals spray of 0.1 per cent carbaryl or fenitrothion is also suggested (Singh & Singh, 1987).

Maharukh (*Ailanthus triphysa*)

Pests : Plantations of soft wood species of *Ailanthus triphysa* are grown around the agricultural field and taungyas. Major pests are lepidopterous defoliators *Atteva fabriciella* (Yponomeutidae) (Plate

Plate 19.4 : a. Larvae of Atteva fabriciella *defoliating* Ailanthus excelsa *b. Moth of* Atteva fabriciella *c. Defoliated taungiya plantation of* Ailanthus excelsa *d. Larval web on* Aidanthus *plants*

19.4) in North and *Eligma narcissus* (Noctuidae) in South. These two pests often are responsible for failure of the plantations. One-two year old toungya plantation of ailanthus raised by seed sowing can completely be wiped off by *A. fabriciella*.

Management: Pests can be managed by integration of biological and limited chemical control methods. Two larval parasites, *Bessa remota* and *Carcelia* spp. (Technidae) and pupal parasite *Brachymeria heimattevae* take appreciable toll of pest population. In nature, many microbial agents like fungi bacteria and nematode operate to reduce the pest population (Mishra, 1978; Jamalludin *et al.*, 1987; Varma 1986; Mathur, *et al.*, 1970). Spraying of synthetic pyrethroid, fenvalerate (Sumicidin) and a carbamate (sevin) in 0.01 and 0.2 per cent concentrations respectively provides good control of the pest (Misra *et al.*, 1987).

Gamhar (*Gmelina arborea*)

Pests : This species is widely grown in taungyas of Assam and North Bengal. Plants suffer from two major pests *viz.*, a defoliator *Calopepla leayana* (Coleoptera : Chrysomlidae) and sap sucker *Tingis beesoni* (Hemiptera : Tingidae). Defoliation by *C. leayana* commences in the beginning of rains and continues till October/November.

Management : Beetles of *C. leayana* are highly attracted to white colour. The control can be achieved by using white light traps when the sun shines after rains. Artificial hibernation and aestivation shelters may be used to trap the insects. It has been observed that attack of beetles mostly occurs on the fringes of a plantation rather than in the centre. This behaviour may be manipulated to control this pest by spraying insecticides on the perifery of plantations (Mathew, 1986). A pupal parasite *Brachymeria excarinata* and a fungus, *Beauveria bassiana* are natural bio-control agents of the beetle (Mohandas, 1986; Sankaran *et al.*, 1989). The control of sap sucker *T. beesoni* can be achieved by applying systematic insecticides *viz.*, monocrotophos (0.03%) or synthetic pyrethroids like cypermethrin or fenvalerate (0.003%).

Toon and Mahagony (*Toona ciliata* and *Swietenia macrophylla*)

Pests : These tree species are generally grown with maize. These grow well under the canopy of secondary growth. A shoot borer, *Hypsipyla robusta* (Lipi : Pyralidae) is a major pests of toon and mahagony plantations. The larvae of *H. robusta* also attack inflorescence and fruits of toon.

Management : Furadan 3 G @ 5 to 25 g per plant can be applied in young crop to control the pest. Sharma (1980) has suggested the use of sex-pheromone for monitoring and suppressing the pest population.

Plate 19.5 : a. Infested babool plant by Celoesterna scabrator *b.* Celoesterna scabrator *larva* in situ *c.* Celoesterna scabrator *beetle d.* Tonica niviferane *moth*

Semul (*Bombax ceiba*)

Pests : The major pests is *Tonica niviferana* (Lepidoptera : Oecophoridae) (Plate 19.5d) which bores the young shoots and twigs of the plants and often cause heavy mortality of plants in young stage.

Management : An ichneumonid *Xanthopimpla brevicauda* is an effective pupal parasite (Chatterjee and Singh, 1969). Rogor (Dimethoate) (0.02%) insecticide may be used to control the pest. Application of granular insecticide carbofuran @ 85 g/plant or sevidol 50 g/plant causes effective mortality of pest in young plantation. Screening of resistant clone of Semul would be useful against shoot borer.

Kadam (*Anthocephalus chinensis*)

Pests : It is extensively grown in West Bengal Agroforestry system. It is attacked by a defoliator *Arthroschista hilaralis* (Lep. - Pyralidae). In August-September pest population reaches at its peak.

Management : Larval parasites *Apanteles balteata* and *Cedria paradoxa* operate well to suppress the pests (Thapa & Bhandari, 1979). Thiodan (0.1%) can be sprayed for the control of Kadam defoliator in young plantations.

Champ (*Michelia champaca*)

Pests : It is grown in northern part of West Bengal in agroforestry system. Champ bug, *Urostylis punctigera* (Hemiptera : Pentatomidae) is a serious pest of this plant species. Sometimes, young plants are also destroyed by the scourge of an aphid *Prociphilus micheliae* which cause heavy curling of leaves.

Management : In nature, the pentatomid bug is controlled by a parasite *Pachyneuron pentatomivora* and a coccinellid predator, *Calvia tricolor* which feeds on egg masses. The bug can be controlled by spraying Phosphomidon (0.2%) (Singh, 1990). Aphids can be controlled by spraying monocrotophos or rogor 0.04 concentrations.

Babool, Khair etc. (*Acacia nilotica, Acacia catechu* and *A.tortilis*)

Pests : These species of *Acacia* are widely planted in agroforestry systems of arid and semi-arid zones. The major pest of *A. nilotica* is a root and stem borer, *Celosterna scabrator* (Coleoptera : Cerambycidae) (Plate 19.5 a, b, c). The ravages of this pest in Central India cause heavy mortality owing to which babul taungyas were abandoned. Besides babul, it also attacks *A. catechu* and *A. tortilis*. It has developed a fancy for *Eucalyptus* spp. also (Chatterjee and Singh, 1979). *Acacia* spp. are also defoliated by bagworm *Cryptothelia crameri*. In South India, another bagworm *Pteroma plagiophleps* is recorded as a pest of *A. nilotica* (Pillai & Gopi, 1990). The cow bug *Oxyrachis tarandus* (Hemiptera; Membracidae) also causes severe damage to young plantations of *Acacias*. Pods and seeds of *Acacia* spp. are severely damaged by bruchid, *Caryedon serratus* (Singh and Bhandari, 1986).

Management : The infestation of *C. scrabator* can be minimized through silvicultural methods. It has collateral hosts like *Eucalyptus*, *Casuarina* etc. These species should not be planted together with *Acacia* spp. For controlling the pest, plantation should be sprayed with 0.2 per cent fenitrothion/endosulfan at the time of emergence of beetles. The application at the base of young plants of furadan 3 G @ 25-50 g/plant at the beginning of monsoons at monthly intervals till October, is also recommended. For control of bagworms spraying of folithion 50 EC (0.25%) or sevin 50 W.P. (0.25%) gives good foliage protection. Sap suckers can be controlled by spraying dimecron or monocrotophos (0.02%). Pod and seed insects can be controlled by spraying -0.25 per cent endosulfan or 0.05 monocrotophos. In stores, seeds of acacias should be dusted with 10 per cent dust of malathion. Fumigation by carbon-disulphide, or aluminium phosphide kills the existing attack of bruchids (Singh and Bhandari, 1986, 1988a).

Eucalyptus (*Eucalyptus* spp.)

Pests : Generally, it is free from defoliator pest in India but occasionally sporadic defoliation is caused by *Suana concolor* (Lepidoptera : Lasiocampidae) in the North. In South India, it is attacked by Phasus borer, *Sahyadrassus malabaricus*. In arid and semiarid region, it is attacked by a root and stem borer, *Closterna scabrator*, which is primarily a pest of *Acacia nilotica*. The major pests of young *Eucalyptus* plants are, however, termites throughout the country.

Management : For the control of termites, use of aldrin, BHC or lindane is suggested in 0.1 to 0.2 per cent concentrations in the soil around the base (Roonwal, 1979; Nair and Verma, 1981). Recently attempt has been made to use less persistent organophosphorous insecticides like chlorophyrifos, phorate, quinolphos etc. against termites in forest nurseries and plantations. Chloropyritos 20 EC in pits at 0.3 per cent concentrations is recommended during transplantation of seedlings (Thakur *et al.*, 1990).

The infestation of *C. scabrator* can be minimized by applying silvicultural methods. Its collatoral hosts like *Acacia* spp., *Casuiarina* etc. should not be planted together with *Eucalyptus*. For controlling the pest, plantation can be sprayed with 0.2 per cent fenitrothion/endosulfan at the time of emergence of beetles. The application of furadan 3 G @ 250 g/plant at the beginning of monsoon at monthly intervals till October is also recommended.

The phassus borer attack is controlled by cutting the plants at the ground level so that a coppice growth can be obtained. Control measures of these pests and other pests have already been mentioned in previous pages.

Poplar (*Populus* spp.)

Poplar is extensively grown in the Tarai area of Uttar Pradesh, Haryana and Punjab.

Pests : Chatterjee *et al.* (1967) listed 60 species causing damage to poplars. But the most dangerous insects are the lepidopterous defolators *Pygaera fulgurita* and *P. cupreata* (Notodontidae). Poplar stem borer *Apriona cinerea* (Coleoptera : Cerambycidae) is also a serious pest of poplar in plantations in U.P. In Kashmir, defoliator *Lymantria obfuscata* (Lept; Lymantridus) defoliate trees (Plate 19.6) and *Aeolesthes sarta* (Coleoptera; Cermabycidae) bores the stem. In Tarai region of U.P. termites *Coptotermes heimi* caused severe damage (Thakur, 1978).

Management : Defoliation epidemics in agroforestry plantation (Taungyas) were controlled by aerial spraying of carbaryl 1.0 kg ai/ha. (Singh *et al.*, 1983). Larvae of *Pygaera* spp. are subjected to the attack of NPV disease. They have been controlled successfully in the field conditions by spraying water suspension containing NPV (Ahmad and Sen-Sarma, 1983). Stem borer, *Apriona cinerea* can be controlled by injecting methyl parathion (0.1%), saturated solution of paradichlorobenzene in kerosene (5 ml/hole) (Verma and Khurana, 1985). The stools of the poplar can be protected by soil application of furadan 3 G @ 5-25 g/plant in July-August. Adults of *A. cinerea* can be destroyed through use of 0.2 per cent Endosulfan spray. In old plants, attacked branches may be prunned before the borer larvae gets entry into the main stem. Defoliator *L. obfuscata* can be controlled by spraying malathion (0.25%) or endosulfan (0.2%) in water (Sen-Sarma and Gupta, 1979).

Plate 19.6 : a. Difoliated agro-forestry plantation of poplar b. Larvae of Pyoaera fulqurita *c. Moths of* P. fulqurita *in copulation d). NPV infested larvae of P. fulgurita d). Moth of* Lymantria obfuscata *with eggs. e.* Apriona cinerea *larva f.* A. cinerea *beetle g. Infested poplar tree by* A. cinerea

Subabul (*Leucaena leucocephala*)

Pests : It was free from any heavy attack of insects till the arrival of a exotic insect *Heteropsylla cubana* (Homoptera: Psylldae). It has threatened subabul cultivation in the country. This insect has been reported from many parts of south and central India and is spreading towards North (Singh and Bhandari, 1988; 1989a; Sivaramakrishnan, 1988). Thakur and Pillai (1985) have reported about 12 native insect pests attacking subabul in South India.

Management: In India, no control measures have been recommended on the basis of experiments. Spray of systemic insecticides like monocrotophos (0.02%) or dimecron is suggested. Use of resistant variety against jumpling lice is advocated in Hawaii and introduction of natural enemies from other countries but recently Singh and Bhandari (1989) warned the introduction of the predators (*Curinus coeruleus*) and *Olla abdominalis* as they may readily predate upon lac insect.

Miscellaneous Tree Species

Prosopis cineraria : It is most popular tree species of arid region of Rajasthan and Haryana. It is also subjected to the attack of many insect species. A termite species *Acanthotermes macroephalus* causes damage to roots in Rajasthan. The beetles of *Holotrichia consanguinea* defoliates the foliage during night. A sap sucker *Oxyrachis tarandus* sucks the sap of tender shoot. Occasionally, stem borer of *Acacia nilotica, Celosterna scabrator* also attack the trees. In Rajasthan and Haryana heavy attack of gall forming chalcid *Trichilogaster* spp. on twigs have been recorded besides leaf gall by eryophyes mite (*Eriophyes prosopides*).

To control the defoliating beetles, spraying of carbaryl 50 WP (0.2%) water emulsion during June-July is recommended. To control the gall forming chalcids propylactic treatment is recommended in September-October when the females are on the wings. Spraying of folithion 0.25 per cent or endosulfan 0.25 per cent is suggested (Singh and Bhandari, 1986a).

Albizia falcateria : It is grown in Kerala in taungyas. The major pest is a bagworm *Pteroma plagiophles* (Lepidoptera: Psychidae).

For controlling this pest, 8 insecticides were screened but quinalphos and methyl parathion (0.05 to 0.1%) were found most effective (Varma, 1986).

Albizia spp. are defoliated by lepidopterous larvae of *Eurema blanda* (Pieridae) and *Laodamia strigvinata* (Pyralidae). *Tamarindus indica* is attacked by *Ophisua janata* (Noctuidae), Haldoo by *Dirades theclata* and cotton leaf roller *Sylepta balteata*, *Celtis australis* is defoliated by beetle *Diorhabda lusca* (Chrysomellidae).

For the management of these defoliators spraying of fenitrothion or Carbaryl (0.1-0.25) is recommended.

Neem (*Azadiracta indica*) : It is a very popular tree of farmers throughout the country. It is attacked by a neem scale, *Pulvinaria maxima* and *Aonidiella* (*Aspidiotus*) *orientalis*. In Neem plantation at Mehsana in Gandhi Nagar, the neem tip borer *Laspeyresia koenigana* was found attacking the neem (Ahmad and Srivastava, 1991).

Scale insects can be controlled by applying systemic granular insecticides like furadan, carbofuran or fodiar spraying of 0.02 per cent monocrotophos rogar etc. For control of neem top borer spray of monocrotophos (0.2%) is recommended (Ahmad and Srivastava, 1991).

Willow (*Salix* spp.)

It is defoliated heavily by Indian gypsy moth *Lymantria obfuscata* (Lymentridae). Besides this, willows are also destroyed by tent caterpiller *Malacosoma indica* (Lasiocampidae) and willow leaf beetle *Plagiodera versicolora*. Giant willow aphid *Tuberolachnus* is a serious pest in Spiti Valley of Himachal Pradesh and Ludakh in J&K.

Gypsy moth, can be controlled by spraying carboryl (0.2%). Tent caterpillar can be controlled by spraying 0.03 per cent endosulfan or monocrotophos (Joshi and Agarwal, 1981). willow aphid can be controlled by spraying 0.04 per cent dimecron.

Mulbury (*Morus alba*)

It is defoliated by *Glyphodes pyloalis* Walker (Lepidoptera: Pyralidae). *Cedria paradoxa* is a promising parasite on it. The pest can be controlled by 0.1 per cent fenitrothion at the beginning of the monsoon. Mulbury is also attacked by borer *Apriona germari* and *Batocera rufomaculata* which can be controlled by injecting saturated solution of paradichlorobenzene in kerosene oil @ 5 ml. into the holes by veterinary syringe and plugging the hole with the wet mud.

Karanj (*Pongamia pinnata*)

It is heavily damaged by a leaf blotch forming lepidopterous, *Lithocolletis virgulata* (Lithocollitidae) and a seed borer *Laspeyresia perfricta* (Torticidae) (Kumar, 1990). Blotch forming insect can be controlled by spraying 0.1 per cent fenitrothion at the time of fresh flush. Seed borer can be controlled by spraying dimecron (0.03%).

Grewia optiva is attacked by defoliators *Lygropia quaternalis* and *Dichocrocis discinotalis*. Defoliator can be controlled by spraying 0.2 per cent per cent fenitrothion.

Insect-host Interaction in Agroforestry System

In agroforestry systems, insect-host interaction is more complex than forest or crop systems. An agricultural or horticultural pest of minor importance may develop fancy for a forestry crop or *vice-versa* and may lead to great disaster. Some of the minor insect pests may assume the status of key pests in the course of time.

There are examples from North-east where the pest of forest trees caused extensive damage to citrus and apple crops. A pest of castor, *Dichocrocis punctiferalis* (Lepidoptera: Pyralidae) begins to defoliates Kashi pine and also attacks fruits of teak (Singh and Singh, 1986). An semi-looper of castor, *Ophiusa janata* (Lepidoptera: Noctuidae) have also developed taste to babul, mahua, imli etc. Some other primarily agricultural pests like gram pod bores *Heliothis armigera* (Lepidoptera: Noctuidae). Bihar-hairy caterpillar *Diacrisia obliqua* (Arctiidae) sometimes heavily damages the forest trees like shisham, teak, subabul, mulbery, toon and cotton bug, *Dysdercus cingulatus* (Hemiptera; Pyrrhocoridae) damages semul and kapok. A few pests of coffee and tea have assumed the pest status of forest trees. Examples may be mentioned of the red borer of coffee *Zeuzera coffea* on teak, albizia and toon (Mathew, 1990). A shoot-hole borer of tea *Xyleborus fornicatus* (Coleoptera: Scolytidae) attacks several species of forest trees including siris, gamhar, silver oak etc. Tea mosquito bug, *Helopeltis antoni* (Hemiptera : Capsidae) that feeds on tea and cashew normally, has developed fancy for neem in South India (Ahmad & Khan, 1988). Besides these there are numerous examples of common food plants of insects in a agroforestry system.

As our forest resources are dipleting at an alarming rate, agroforestry practice is being adopted by farmers on a large scale to fulfil their needs of fuel, fodder and small timber. In such a changed multicropping habitat, new types of insect problems may arise in future and may pose serious challenges to the forest and agricultural entomologists. Their job will therefore, be more complex and challenging than ever before. In such a agroforestry system, coordination and cooperation of agricultural and forest entomologists are required to great extent to investigate biology, population dynamics and strategies for control of agroforestry pests. Further, pests of fodder trees cannot be controlled by insecticides and needs biological and cultural control measures.

Conclusions

It has been realised that an agroforestry ecosystem is an intermediate stage between agro-ecosystem and sylvan ecosystem. It is unstable in comparison to forest ecosystem, where a balance between the insects and its enemies exists to maintain pest population at endemic. In such an unstable ecosystem, a low profile insect of a forest or agricultural crop may assume the high profile status owing to ample food supply throughout the year and lack of natural enemies. The greatest danger in such a system is the attractives of an agricultural pest towards forest crop for its food supply or *vice versa*. Such change in food habit may brought the development of a mono or oligophagous insect species into a polyphagous species which may cause disaster to both the crops.

The management of agroforestry insects damaging forest crops can be taken up together with the management of agricultural crop pests. The use of insecticides should be restricted. Foliage of fodder trees of agroforestry system should be protected by such insecticide which has very low residual effect e.g. pyrathrins etc. Biological and cultural control is the best way to manage the agroforestry pest problems.

Future Research Needs

In India, negligible work has been done on this fruitful line of research. This field requires immediate attention of entomologists to investigate the occurrence and population dynamics of pests in agroforestry systems so as to evaluate the safe combination of forest tree species with agricultural crops. Sometimes, forest nurseries have to be established in agricultural dominated regions. These nurseries are most vulnerable from the attack of agricultural pest of adjoining areas. Therefore, a greater coordination between the forest and agricultural entomologists is needed to have a joint front against the tiny but very powerful enemy, "insects".

REFERENCES

Agarwal, J.P. Rajak, R.C., Katre Purnima and Sandhu, S.S. (1985). Studies on entomophazons fungi parasitizing insect pests of teak. *J. Tropical Forestry*, 1(1), pp. 91-94.

Ahmad, I. and Khan, Salar (1988). *Helopeltis antonii* signoret (Miridae : Heteroptera), a potential pest damaging neem trees in Tamil Nadu with a special note on its control. Abstracted in *Proceedings of 75th Indian Science Congures, Pune*. Part IV, pp. 115-116.

Ahmad, S.I. and Sen-Sarma, P.K. (1983). Investigation on a newly recorded nuclear polyhedrosis virus of *Pygaera fulgurita* Walk (Notodontidae : Lepidoptera) I. Nature of disease and insect virus interactions. *J. Ent. Research* 7(1), pp. 150-153.

Ahmad, I. and Srivastava, K.K. (1991). A report on the insect pest and disease problems in the forest nurseries and plantations of Gandhinagar and Meshana Forest Divisions (Gujarat) with possibilities of their management. Tour Report Institute of Arid Zone Forestry Research, 17 pp + Tables and Plates.

Ahmad, M. (1987). Relative resistance of different clones of *Tectona grandis* to teak defoliator *Hyblaea puera* Cram. (Lepidoptera : Hyblaeidae) in South India. *Indian Forester, 113*(4), pp. 281-286.

Beeson, C.F.C. (1941). *Ecology and control of forest insects of India and adjacent countries*. Vasant Press, Dehra Dun 1007 pp.

Beeson, C.F.C. and Chatterjee, S.N. (1939). Further notes on the biology of teak defoliators in India. *Indian Forest Record* (N.S.) Ent. 5(5), pp. 357-359.

Bhasin, G.D. and Roonwal, M.L. (1954). A list of insect pests of forest plants in India and the adjacent countries. Parts 1 & 2. Indian Forest Bulletin (New Series): 171-(1).

Chatterjee, P.N. and Singh, P. (1968). *Celosterna scarabrator* Fab. (Lamiidae : Coleoptera), a new pest of *Eucalyptus* and its control. *Indian Forester* 94(11) pp. 826-830.

Chatterjee, P.N. and Singh, P. (1969). Biological control of semul shoot borer *Tonica niviferana* walker (Oecophoridae). Biology of *Xanthopimpla previcauda* (Cushman) (Ichmeumonidae) a pupal parasite of *Tonica niviferana* Walker. *Indian Forester*, pp. 95.

Chatterjee, P.N., Singh, P. and Shivaramakrishnan (1967). A list of insect pests of *Eucalyptus nurseries* and plantations. *Proceeding All India Symp. on Eucalyptus, Poplars and Willows, Dehra Dun*, pp. 1-5 cyclostyled.

Jamaluddin, Sahare Sima, Mesh Ram, P.B. and Dadwal, V.S. (1987). A note on pararitism of *Aspergillus flaieus* on *Ailanthus* webworm. *Indian For.*, 113(10), pp. 707-708.

Joshi, K.C. and Agarwal, S.B. (1981). Efficacy of some insecticides against the larvae of tent caterpillar, *Malacosoma indica* Wilk., *Indian J. Agric. Res.*, 15(4), pp. 251-257.

Kedarnath, S. and Singh, P. (1975). Studies on natural variation in susceptibility of *Tectona* to leaf skeletonizor *Pyrausta machaeralis* (Lepidoptera : Pyralidae). *Proc. FAO/IUFRO Symposium on Forest Diseases and Insects*, New Delhi, pp. 1-7.

Khosla, P.K., Puri, Sunil and Khurana, D.K. (1985) (Eds.) Agroforestry systems - A new Challenge, ISTS. Solan (HP), pp. 1-300.

Kumar, Sushil (1990). Insect pests of some socio-economic tree species of tribal areas in Madhya Pradesh and their control. In : *Social forestry in tribal regions and environmental management* (Eds. R.N. Trivedi, P.K. Sen-Sarma and M.P. Singh) 5, pp. 53-67. (Today and Tomorrows Publication, New Delhi).

Kumar, Sushil and Thakur, M.L. (1989). Damage to nursery stock by a rodent *Nesokia indica* (Gray) at Satyanarayan Forest Nursery, Dehra Dun (Uttar Pradesh). *Indian Forester* 115(3), pp. 177-179.

Lal, J.B. (1989). *Indian Forests - Myth and Reality*. Natraj Publishers, Dehra Dun, pp. 304.

Mathew, G. (1986). Insects associated with forest plantations of *Gmelina arobrea Roxb*. in Kerala, India *Evergreen*. 9(4), pp. 308-311.

Mathew, G. (1990). Cossid pests of teak in the Asian Region and the possibilities of their control. *Proceeding IUFRO Regional Workshop on Pest and Disease of Forest Plantations*, pp. 204-208.

Mathew, G. and Ali, M. (1987). Microbial pathogen causing mortality in the carpenter worm, *Cossus cadambae* Moore (Lipidoptera : Cossidae), a pest of teak (*Tectona grandis*) Linn. f. in Kerala (India). *J. Tropical Forestry* 3(4), pp. 349-351.

Mathur, R.N. (1959). Mermis sp. (Mermithidae, Ascaroidea - Nematoda) and its insect hosts. *Current Science*, 28, pp. 255-256.

Mathur, R.N., Chatterjee, P.N. and Sen-Sarma, P.K. (1970). Biology, Ecology and control of *Ailanthus* defoliator *Atteva fabriciella* Sweed. (Lipidoptera : Yponomeutidae) in Madhya Pradesh. *Indian Forester 96*(7), pp. 538-552.

Meshram, P.B., Pathak, S.C. and Jammaluddin (1990). Effect of some soil insecticides in controlling the major insect pests in teak nursery. *Indian For. 116*(3), pp. 206-213.

Misra, R.M. (1978). A mermithid parasite of *Atteva fabriciella* Sweed. *Indian For. 104*(2), pp. 133-134.

Misra, R.M., Prasad, Ganga and Rawat, D.S. (1987). Control of *Ailanthus*, webworm *Atteva fabriciella* Swed by Chemical Insecticides in Plantations. *Indian For. 113*(2), pp. 147-149.

Mohandas, K. (1986). *Brachymeria excarinata* Gahan (Hymenoptera : Chalcidae) as pupal parasitoid of *Calopepla leayana* Latr. in Kerala, India, a new record, *Entomon, 11*(No.4), pp. 279-280.

Nair, K.S.S. (1982). Seasonal incidence, host range and control of the teak sapling borer *Sahyadrassus malabaricus*, K.F.R.I. Report No. 16.

Nair, K.S.S. and Varma, R.V. (1981). Termite Control in *Eucalyptus* plantations. KFRI Research Report, No. 6, pp. 48-49.

Neelay, V.R., Bhandari, R.S. and Negi, K.S. (1983). Effect of insecticidal and hormonal spray on the production of fruits in teak seed orchards. *Indian For.* 109(11), pp. 829-839.

Patil, B.V. and Thontadarya, T.S. (1981). Record of *Beauveria bassiana* (Balsamo) Vuilemin on teak skeltonizer *Pyrausta macheralis* (Walker). *Indian For.*,107, pp. 698-699.

Patil, B. and Thontadarya, T.S. (1983a). Natural energy complex of the teak skeletonizer *Pyrausta mechaeralis* (Walker) (Lepidoptera : Pyralidae) in Karnataka, *Entomon*, 8(3), pp. 249-255.

Patil, B.V. and Thontandarya, T.S. (1983b). Studies on the acceptance and biology of different *Trichograma* spp. on the teak skeletonizer *Pyrausta mechaeralis* Walker, *Indian For.* 109(5), pp. 292-297.

Pillai, M.S.R. and Gopi, K.C. (1990). The bagworm *Pteromaplagiophleps*

Hamp (Lepioptera Psychidae) attack on *Acacia nilotica* (Linn.) Wild ex.del. *Indian For.* 116(7), pp. 581-583.

Roonwal, M.L. (1979). Termite life and termite control in tropical, South Asia, XII+177 pp.

Sen-Sarma, P.K. (1986). Some important problems of forest insect pests and Research Programmes. *Proceeding 2nd Forestry Conference*, 1980, Vol. II, pp. 879-883.

Sankaran, K.V., Mohandas, K. and Ali, M.I. (1989). *Beauveria bassiana* (Bals) Vul. a possible bio-control against *Myllocerus viridanus* Fabr. and *Calopepla leayana* Latreille in South India. *Current Science* 58(8), pp. 467-469.

Sen-Sarma, P.K. and Gupta, B.K. (1979). Insect pests of Poplars. *Symposium Proceedings Silvicultural Management and Utilisation of Poplars, Srinagar* (Ed. R.V. Singh), pp. 121-122.

Sharma, K.K. (1980). Sex pheromones of Meliaceae shoot borer *Hypsipyla robusta* (Lepidoptera : Pyralidae) and investigation into the possibilities of their use in survey and suppression traps. Ph.D. Thesis submitted to Garhwal University, Srinagar, Pauri Garhwal (U.P.).

Singh, G., Arora, Y.K., Narain, P. and Grewal, S.S. (1990). *Agroforestry Research in India and other Countries.* Surya Publication, 4-B, Nashville Road, Dehra Dun (U.P.) pp. 187.

Singh, P. (1990). Insect pests in plantations of native tree species in India, IUFRO Workshop on pests and Disease of Forest Plantations Bankok, pp. 45-55.

Singh, P. and Bhandari, R.S. (1986). Insect pests of forest tree seeds and their control. *Proc. National Seminar on Forest Tree Seed*, pp. 155-171.

Singh, P. and Bhandari, R.S. (1986a). Insect Pests of *Prosopis* and their control. *Proc. Nat. Seminar on Role Prosopis in Wasteland Development*, Bhavnagar (1986) pp. 1-8 (Photocopy).

Singh, P. and Bhandari, R.S. (1988). The arrival of the *Leucaena* psyllid in India. *Leucaena Res. Rep.*, 9(1), page. 20.

Singh, P. and Bhandari R.S. (1988a). Insect pests of leguminous Forest tree seeds and their control. *Indian For.*, 114(12), pp. 844-853.

Singh, P. and Bhandari, R.S. (1988b). Insect pests of bamboo and their control. *Indian For.*, 114(10), pp. 670-683.

Singh, P. and Bhandari, R.S. (1989). Biological control of *Leucaena* psyllid *Heteropsylla cubana* Craw. (Hemiptera : Psyllidae) - a word of caution. *Leucaena Res. Rep.* (Hawaii) 10, pp. 10.

Singh, P. and Bhandari, R.S. (1989a). Further spread of *Leucaena* psyllid *Heteropsylla cubana* in India. *Indian For.* 115(5), pp. 303-309.

Singh, P. and Bhandari, R.S. (1989b). Insect pest problems in forests and plantations. Meeting of F.R.I. Scientist with Forest Officers of Haryana, Kalesar, April, 1988, pp. 1-10.

Singh, P. and Gupta, B.K. (1978). Laboratory evaluation to inisecticides as contact spray against forest pests. (Lepidoptera : Pyralidae). *Indian Forester.* 104(5), pp. 359-366.

Singh, P. and Misra, R.M. (1978). Bioassay of thuricide - a microbial insecticide against important forest pests. *Indian For.*, 104(12), pp. 838-842.

Singh, P. and Misra, R.M. (1987). New Record of *Beauveria tenella* (Delacroix) Siemaszka on teak Skeletonizer *Eutetona machaeralis* Walk. (Lepidoptera : Pyralidae). *Indian For.*, 113(7), pp. 476-478.

Singh, P. and Singh, R. (1978). Two new pests of Khasi pine *Pinus kesiya* Royle Ex Gordon. *Indian For.*, 104(3), pp. 182.

Singh, P., Rawat, D.S., Misra, R.M., Fasih, M., Prasad, G. and Tyagi, B.D. age S. (1983). Epidemic defoliation of Poplars and its control inTarai Central Forest Division, Uttar Pradesh, *Indian For.*, 109(9), pp. 675-693.

Singh, P. and Singh, S. (1987). Pest and pathogen management in agroforestry systems In: Proceeding of agroforestry for rural needs. Vol. I. (Eds. P.K. Khosla and D.K. Khurana) pp. 153-177 (ISTS Solan).

Singh, P. and Thapa, R.S. (1988). Defoliation epidemic of *Ascotis selenaria imparta* Walk (Lepidoptera : Geometridae) in sal forest of Asharori Range, West Dehra Dun Forest Division. *Indian For.* 114(5), pp. 269-274.

Sivaramakrishanan, V.R. (1988). Arrival of the Jumping lice of *Leucaena* into Karnataka and suggestions on control. *Silva News Letter* No. 167, Bangalore.

Sudheendra Kumar, V.V. (1986). Studies on the natural enemies of the teak pests *Hyblaea puera* and *Eutectona machaeralis*, K.F.R.I. Research Report No. 38.

Thakur, M.L. (1978). The problem of termite damage in poplars in the Bhabar Tarai region of Uttar Pradesh. *Indian J. Forestry*, 1(3), pp. 217-222.

Thakur, M.L. and Sen-Sarma, P.K. (1980). Current status of termites as pests of forest nursuries and plantations in India. *2nd Forestry Conference* 1980, Vol. II, pp. 883-887.

Thakur, M.L., Kumar, Sushil, Negi, A., Rawat, D.S. and Rai, A. (1989). Chemical control of white grubs in forest nursuries under arid soils. *Tropical J. Forestry* (in press).

Thakur, M.L., Sharma, S.D., Prasad, G.K., Kumar Sushil and Negi, A. (1990). Efficacy of some insecticides in the control of termites in *Eucalyptus* plantation under sodic soil. *Proc. National Seminar on Technology for Afforestation Wasteland*, November, 1990.

Thakur, M.L. and Pillai, S.R.M. (1985). Insect fauna of Subabul (*Leucaena leucocephala* (Lann. De Wit) from South India. *Indian For.*, 111(2), pp. 68-77.

Thapa, R.S. and Bhandari R.S. (1976). Biology, ecology and control of Kadam defoliator, *Arthroschista hilaralis* Walk. (Pyrahdae, Lepidoptera) in plantations in West Bengal. *Indian Forester* 102(6), pp. 388-401.

Vaishampayan, S.M. and Bhandari, R.S. (1981). Chemical control of white grub *Holotrichia consanguinea* Blanch (Scarabaeidae : Melolonthinae) in teak nursuries. *Pestology* 8, pp. 5-10.

Verma, R.V. (1986). Seasonal incidence and possible control of important insect pests in plantations of *Ailanthus triphysa KFRI Research Report* No. 39, pp. 1-42.

Verma, T.D. and Khurana, D.A. (1985). Bionomics and control of poplar stem bores, *Apriona cinerea* Chevrolat (Lamiidae : Coleoptera) in Agroforestry Plantations. In : Agroforestry Systems - a new challenge (Eds. Khosla *et al.*,), pp. 237-240.

Chapter 20

Forestry Proposals to Improve the Socio-Economic Condition of the Tribals of Chotanagpur and Santhal Parganas

Lalit Jha and Punam Roy***

A large number of tribals earn their livelihood by selling wood for fuel and by collecting and selling minor forest produce. Villagers are facing problems in collection of firewood and the tribal women have to trudge an average distance of 5 to 10 kms and spend more than half day for only a head load of 5 to 10 kg. It is estimated that per family requirements of fuel, fodder and timber is 6 kg, 5 kg and 5 cft respectively. Farmers can adopt Bund plantation, Nock plantation, Agri-Silviculture, Agri-Silvi-Horticulture or Agri-Horticulture to fulfil their domestic requirement of fuelwood.

The only visible solution is by closely involving the rural people to utilise all land available to plant trees. Single purpose primary society, namely Social-Forestry Co-operative Society may be formed, which will assist in the speedy implementation of this programme.

The Bihar State, has a total geographical area of 173867 sq. km. The Chotanagpur plateau is hilly and is characterised by forests, mines, tribals and other weaker sections. It covers about 46 per cent of the total geographical area of Bihar.

Districtwise Distribution of Tribals in Chotanagpur and Santhal Pargana

As per 1981 census, the tribal population in the State of Bihar was around 5.8 million. About 92 per cent of the tribal people resides in

**Regent tower, 3A, Netaji Subhas Road, Calcutta.*

***K.N. Jha, Vill + P.O. - Karnpur, Dt-Supaul, Saharsa (Bihar).*

Chotanagpur and Santhal Parganas, where the distribution of tribal population is as follows (Srivastava, 1985) :-

Old Ranchi district	56.4%
Singhbhum	44%
Old Santhal Parganas district	36.7%
Palamu	18.3%
Giridih	12.9%
Dhanbad	9.1%
Hazaribagh	9%

Tribals Vs. Forests

The rural areas of plateau region are inhabited by tribals and other weaker sections.

A large number of tribal earn their livelihood by selling fuelwood and various minor forest produce.

A diagnostic survey was carried out in Chotanagpur and Santhal Parganas for conducting this study. 50-60 farmers from each category of land holding i.e. up to 2.5 acres, 2.5 to 10 acres and more than 10 acres were sampled. In Chotanagpur and Santhal Parganas fuelwood, animal dung and agricultural waste provides 95 per cent of the energy requirement of rural populace. Villagers are facing problems in collection of firewood and the tribal women have to trudge an average distance of 5 to 10 km. and spend more than half a day for only a head load of firewood. It is estimated that the per family consumption of all fuel (wood, agriculture wastes, dungs, etc.) put together is more than 6 kg per day. Reasons behind this is nonavailability of other alternative source of energy like cooking gas, electricity etc. Biogas has not been popularised. It is apparent that generally villagers are not using cowdung as manure. They use it for cooking purpose (Jha, 1989).

In Chotanagpur and Santhal Pargana live-stock constitutes an important component of the economy of villagers. The economic return is very small, while damage caused by such a large live stock population to the vegetative cover and environment is very serious. It was difficult to estimate the quantity of fodder required for each cattle but as per villagers' information about a head load grass (20-25 kg.) is required for 5 cattle per day i.e. on an average of 5 kgs per cattle per day.

As per estimation (based on survey) conducted by staff members of AICRP Agro-Forestry, B.A.U. centre, on an average 4 to 5 cft timber are utilised by each household for repair and replacement annually.

Forestry Proposals for Chotanagpur and Santhal Pargana

Proposals discussed and below will definitely solve the requirements of the rural tribal women of this area, as well as the problem of unemployment. Most of the operations may be carried by the tribal women.

Proposal-I

Plantation on Bunds

In these areas, women spend 40 to 50 per cent of their energy in procuring fuelwood. It clearly indicates that more than 150 working days of the year are needed to ensure sufficient firewood for one family. The working hours spent on collecting wood cannot thus be used for work in the fields.

The only visible solution is by closely involving the rural people to utilise all land available to plant trees. Bagchi (1986) has stated that any wood which can be burnt is not fire wood. The provisional definition of firewood is the wood which grows at a faster rate, which otherwise cannot be used as good timber and produces greater/kg/calories of heat per unit weight. It is therefore essential to develope suitable strains and varieties of trees best adopted to different agro-climatic conditions.

In this area we should select such types of tree species which are characterised by faster growth rate, good coppicing ability and greater fuel efficiency.

Presuming that one hectare of land has 600 metres of farm bund, the farmers shall require 200 seedlings if planted at a spacing of 3 metres apart, all along bund in one row. In this area two rows may be planted. Number of seedling for two rows will be 400.

The species proposed for bund plantation :-

1. *Eucalyptus teriticornis*
2. *Dalberia sisoo*
3. *Albizia labbek*
4. *Leucaena leucocephala*

Benefits

The following benefits may be expected after adopting bund plantation:

The sale value of sisoo trees alone after 8th year will be nearly Rs. 75 per tree. So the total price of 200 trees (in one row) will be 200 X 75 = Rs. 15000.00

The expenditure incurred in planting operations as well as protection will be Rs. 2000.00 for three years. So the profit will be approximately Rs. 13,000.00. Farmers can collect small timber etc. after 5th year. This timber may be utilised as timber for hutments, agricultural implements or as fuelwood. After 10 to 15 years farmers can collect good timber from sisoo. Farmers can sell it in the market this will boost up farmers economic condition.

Besides this, farmers can plant *Leucaena leucocephala*. Farmers will get fodder to feed cattle and also maintain a milking cattle. Eucalyptus is a good fuelwood and very fast growing Eucalyptus poles after 5 years can fetch Rs. 200 per tree.

In general one can say that after adopting bund plantation the socio-economic condition of farmers shall improve.

Proposal-II

Farm Forestry

Farm forestry advocates growing of trees which would yield small timber, firewood, fodder, green manure, oil seeds, fruits and other minor forest products, so that the farmers become either self-sufficient, or get supplementary income by selling any excess quantity of any of the above forest products after reserving certain quantities for his own use. The bigger farmers can set at least 10 per cent of the farm land for growing economically important trees. They can in addition, utilise all lands unfit for agriculture and horticulture for growing useful forest species (Rajan, 1986).

It is apparent from Table 20.1 that if sisoo is planted at 2 x 2 meters spacing in one hectare, we plant 2500 seedlings. Farmers can sell these trees after 8th years. Farmers will get easily Rs. 75 per tree. The total price of 2500 trees will be = 2500 x 75 = Rs. 1,87,000 (approx) after 8th year in one hec plantation. Total expenditure upto 8 years will not be more than Rs. 10,000/- per hec.

Table 20.1 : (Economics of sissoo plantation on one hectare of land (spacing 2 x 2 meters, Total number of plants 2500) (in Rupees).

Sl. No.	Item	Years 1st	2nd	3rd	4th	5th	6th	7th	8th
A.	*Labour Cost*								
1.	Land preparation, pit digging etc.	70							
2.	Fencing	40		10					
3.	Transportation of seedlings	6							
4.	Plantation and casuality replacement	30	5	2	1				
5.	Watering	40	30						
6.	Watch and Ward (can watch 12 hec. land)	30	30	30	30	30	30	30	30
Total labour @ 15/day		3240	975	480	465	450	450	450	450

Table 20.1 - Contd.

Sl. No.	*Item*	*Years*	*1st*	*2nd*	*3rd*	*4th*	*5th*	*6th*	*7th*	*8th*
B. *Material Cost*										
1.	Fencing materials		400	-	-	-	-	-	-	-
2.	Manure Insecticide treated FYM or Urea		300	-	400	-	-	-	-	-
3.	Cost of 2500 *Dalbergia Sisoo* seedling @ Rs. 0.50		1200	-	-	-	-	-	-	-
	Total Material Cost		1,900	-	400					
	Total labour + Material cost		5,140	1,080	880	465	450	450	450	450
C. *Returns*										
1.	Sale of branches									
	@ 5 kg. per tree, lopped branches can be sold @ Rs. 2									
	Sale of timber @ 75 per tree									1,87,500
	Total Return		--	-	-	-	-	25,000		1,87,500
			(-)	(-)	(-)	(-)	(-)	(+)	(-)	(+)
	Net Return		5,140	1,080	880	456	450	24,450	450	1,87,500

Table 20.2 : Economics of Mango plantation on one hectare of land

Items	*1*	*2*	*3*	*4*	*5*	*6*	*7*	*8*	*9*	*10*	*11*	*12*	*13*	*14*	*15*
A. *Labour cost*															
(i) Survey, Demarcation -15 lay out	15														
(ii) Fencing-50	50			20		10									
(iii) Pit digging	20														
(iv) Transport of seedling	6														
(v) Plantation	10														
(vi) Casuality replacement	2	5	2	1											
(vii) Hoeing & Weeding	10	10	10												
(viii) Watering	40	40	30					40							
(ix) Watch & ward in fruitning season				30	30	30	30	30	30	30	30	30	30	30	30
(x) Harvesting				2	2	2	3	4	4	4	5	5	5	5	5
Total labour	153	55	42	20	32	42	32	74	34	34	35	35	35	35	35
@ 15/day	2295	630	300	380	480	630	495	1110	510	510	525	525	525	525	525

Table 20.2— Conted.

Items	*1*	*2*	*3*	*4*	*5*	*6*	*7*	*8*	*9*	*10*	*11*	*12*	*13*	*14*	*15*
B. *Material Cost*															
(i) Fencing materials	400														
(ii) Manure	300	300	300	400	400	400	500	500				600			700
(iii) Supling @ Rs. 10	1000	200	40												
Total	1700	500	340	400	400	400	500	500				600			700
Material Cost															
Total Labour	3,995	1,325	970	700	880	1,030	995	1,610	510	510	525	1,125	525	525	8,200
Material cost															
C. *Return*															
(i) Fruit sale					1000	4000	5000	6500	7500	7500	7500	7500	8000	8200	8200
(ii) Sale of timber/fuel 400 tonnes @ Rs. 325 per/tonnes															1,30,000
Total return					1000	4000	5000	6500	7500	7600	7600	7600	8200	8200	8200
Net return	(-) 3995	(-) 1325	(-) 970	(-) 700	(+) 120	(+) 2970	(+) 4005	(+) 4890	(+) 6990	(+) 6990	(+) 7075	(+) 6475	(+) 7475	(+) 7675	(+) 7675

Similarly Block plantation of any fruit tree (e.g. mango) will be much economical (Table-20.2).

In Kumharia village adopted under Lab to Land programme by the Faculty of Forestry, Birsa Agricultural University. 75 per cent villagers are interested to bring their land under tree cover to supplement their agricultural income. This clearly indicates healthy signs of the villagers and role of Lab to Land Society Forestry programme in motivating the villagers to adopt Social Forestry programme on their lands.

Choice of species: *Dalberia sisoo, Eucalyptus tereticornis Terminalia arjuna*, Mulberry *Albizia labbek* etc.

Jha (1985) has stated that Arjun may be planted in close spacing of 1.2 x 1.2 or 1.5 m for rapid promotion of Tassar culture. In tassar culture poor people of tribal areas get employment in the following ways :-

(i) Employment potential is generated in the collection and preservation of superior quality seeds of plants, sowing of seeds and watering seedlings.

(ii) For the preparation of pits, planting and polarding of trees.

(iii) Rearing of tassar silk worms and collection of cocoon provide job to the weaker sections of our society.

In brief, it may be stated that tassar culture is a labour intensive programme and is helpful particularly to the weaker section in tribal areas. Prasad and Jha (1985) have reported that it will be more profitable if some spinning machines are provided to the tribals. Many tribal boys and girls are already engaged in this trade. These youngsters earn Rs. 15 to 20 per day and remain in their own village.

(iv) Similarly Block Plantations of mulberry will be helpful in rearing mulberry silk and member of the family right from adult to minor will get opportunity to work and earn good income.

(v) Farmer can adopt block plantation for fuelwood trees. As we know, fuelwood, animal dung and agricultural waste provides 80 per cent source of energy in rural areas. In most of the villages people are in the habit of meeting firewood requirements from the near by forests. Block plantation of fuel wood under farm forestry will improve rural economy and meet the requirements of fuelwood and this will indirectly provide chance to the traditional forests to develop.

Proposal-III

Village woodlots

In order to reduce pressure on forests and to given people forest resources in the vicinity of their own villages, community wastelands can be taken for plantation with an agreement between the social forestry department and the village Panchayats for the protection and management of these plantation and sharing of benefits.

Plantation in the community land will definitely stop collection of headloads from forests in which damage is mainly caused to pole crop and younger regeneration.

Species recommended for plantation under farm forestry may be used under this scheme.

In villages, woodlots planting should by done by the forest department who will maintain these plantations for the first few years which should be handed over to the villagers later.

Waddin (1987) has suggested that be providing such type of woodlots pressure on traditional forest will be reduced.

Benefits

- When management will be handed over to the village panchayat, the villagers will remove grass, fallen wood, loppings from these plantation free of cost.
- The harvested material will be utilised by farmers of village, to meet their requirement of fuel wood etc.

Proposal-IV

Pasture Land

In Chotanagpur plateau, barren land may be developed to pasture land. This land, for the past many years due to over grazing has become compact and demanded and does not support adequate much grass cover to provide fodder to the cattles.

(a) Silvi-Pasture System Silvi-pastural practices can ensure high yields of good quality grass and fodder besides maintaining the productivity of such sites. The trees tap the lower layers of the soil which will otherwise

not be tapped by grasses. Silvi-pastural practices will thus ensure better utilisation of the land.

Singh (1984) has stated that under rainfed conditions improved grass adapted to the area should be developed or sown as a two or three tire system. Here selection of trees could be either for timber alone or for dual purpose i.e. for wood and fodder.

I. *Two tire system :* The species selected herein are based on economic needs of the villagers. Once the trees have established themselves and attained reasonable growth, highly nutritive perennial grass seeds are either broadcast or sown as an understorey crop.

The following species can by sown/planted as understorey crop. These crops do not face competition for solar light among themselves, and are not affected by shade and by the root exudates, if any, of trees.

Grass species	-	Dinanath grass, Sada Bahar
Tree species	-	*Dalberjia sissoo, Albizia lebbek Leucina leucocephala* etc.

II. *Three-tire system*: Singh (1984) has suggested that this subfacet involves incorporation of another component i.e. pasture legumes besides the grasses as inter-crop among the trees. These under storeyed legumes are, by and large, short statured perennial and somewhat spreading in habit and thus they occupy the lower most tier. The three combinations are based on the principle that each of its component draw nutrients from different layers of soil and are complementary to each others growth rather than competitive.

The most productive legumes found compatable with grasses are *Stylosanthis quianensis*, *Macroptilium atropurpureum* and *Clitorea ternatea.*

The advantages of growing grasses and legumes among the fodder trees are that during the wet season or the period in which these forage species remain, green, animals look to the ground for green foliage but during the lean period animals can look up for the tree foliage as top feed for sustenance.

(b) Pastural land can easily be developed in Chotanagpur Plateau by simply closing the area for 5 years from grazing and introducing good palatable grass. The area should be fenced by trench fencing. In such area, tees of miscellaneous species may be planted and in the interspace improved grass may be broadcasted and area may be devided into 4 parts. From 2nd years, villagers may be permitted to cut grass free of cost for their bonafide use only. This type of demonstration plot was established in village Kumharia under Lab to Land Programme which motivated the beneficiary villagers to adopt the technology.

Benefits

- If all the pasture area will be tacked the yield will be multiplied every year and full requirement of grass, fodder is met.
- Villagers may also collect small timber for domestic energy needs.
- Grasses conserve soil and moisture. Thus, improving physio-chemical properties of the soil.

Agroforestry

As is well known, Agroforestry implies cultivation of Agricultural crops in conjunction with tree crops in the same unit of land.

This system would be very useful for farmers in improving their living standards. Agroforestry thus provides a big scope in not only increasing production of wood but also promoting increased production of cereals, fodder, oilseeds and legumes. Experiments have shown that selection of suitable species which do not compete with the field crops can help in increasing the yield of the field crops.

Tree species suitable for villagers under Agroforestry System

Sissoo, Gamhar, Albizzia, Cassia, Bakain, Neem, Babul, Khair, and other Acacias and Arjun etc.

Benefits

Main benefits of Agroforestry programme are as follows :-

1. It will increase the supply of fuelwood in the village at a convenient distance and thus meet the growing requirements of fuelwood of the society. In India, it has been estimated by the experts that rural population use dried dung approximately 70 million metric tonnes and thus the supply of organic (natural) manures for agriculture is reduced tremendously. Through promotion of Agroforestry, the supply of fuelwood

in rural areas can be raised adequately and cowdung can increase the soil fertility. Moreover, the undue pressures put on traditional forests for fuelwood for the inhabitants of adjoining villagers, will be reduced to a large extent.

2. Agroforestry aims to raise the supply of small timbers used by the villagers for agricultural implements, house construction and some other domestic purposes. Agroforestry can thus, meet these requirements of villagers and reduce pressures on traditional forests.

3. One of the main advantage of this project is to raise the production of food crops/legumes/tubers/fruits/vegetables to meet the nutritional stability of the villagers.

Agroforestry programme is labour intensive which thus creates new employment opportunities among the landless labourers and small marginal cultivators. Their income may be appreciably enhanced by this programme thus promoting harmonious Social system in the village.

Agroforestry programme also helps in creating ecological balance in rural areas which is a matter of great significance for a country like India.

This programme will help in the conserving moisture in culturable land and check soil erosion. In this way it may be helpful in raising the produce of the land.

In drought prone condition, Agroforestry projects reduce the insecurity of the agriculturists. It has been claimed that in such areas the dual system of production of tree and grasses of legumes etc. tends to stabilise productivity of the land.

In short, this programme will help in raising the total yield of land considerably in comparison to traditional system of land management where dichotomy between forestry and agriculture/horticulture/animal husbandry operates.

Jawahar Rozgar Yojana and Social Forestry

The most important aspect of this Yojana is that preference shall be given to SCs/STs for employment and 30 per cent of the employment opportunities under the Yojana will be reserved for the women. Social Forestry has been given a place of pride in the JRY. There is a provision for financial assistance for Social Forestry work under JRY. In the JRY it is clear that higher financial assistance for farm forestry will be given to Scheduled castes and Scheduled tribe beneficiaries.

Wherever tree plantation has been taken up under the Social Forestry Programme on community land, cost of maintenance of such plantation upto the three years can be met from JRY funds.

However, under JRY there is no provision of funds for protection of individual farmers land against grazing, human vandalism etc.

Suggestions

Formation of Social Farm Forestry Co-operative Society

Tables 20.1 and 20.2 indicate that in forestry operation financial assistance is essential otherwise farmers will not take interest in any type of plantation scheme, because they can't get income in same year through tree the plantation operation. Hence, long term financial assistance and other types of incentives are needed, so that resource starved tribal farmers can actively participate in above project.

On the basis of the recommendation of K.S. Bawa, large sized agricultural multipurpose societies were formed through out the tribal belt of the country.

Presently utmost emphasis is laid for the promotion of Social Forestry and hence single purpose primary society, namely Social-Forestry Co-operative Society may be formed in Chotanagpur and Santhal Parganas which will help in the speedy implementation of Social Forestry programme, which will reduce economic exploitation and raise the socio-economic condition of tribal women and men. Under Social Forestry Co-operative society -

(i) Each member shall be provided timely short, medium and long term credit for consumption purpose to start above project work and also during the lean period of the year to eliminate exploitation from money lenders.

(ii) Inputs of superior quality planting material shall be supplied to the members at a concessional rate without delay.

(iii) Proper technical guidance shall be provided by inviting experts from Forest Deptt. or Forestry Faculty of the University.

(iv) Marketing of produce of members shall be exclusively done by primaries, so as to eliminate exploitation by the intermediaries.

(v) Processing and marketing of the produce of the members by the primaries will facilitate the timely recovery of loan and hence the problem of outstanding loans will be automatically checked.

Social Forestry Co-operative Societies may be formed in suitable circumstances. In such primaries, the members have to pool their individual land and consequently the production work is to be managed by the executive committee of the societies.

A cell for the promotion of Social Forestry in the Department of co-operative under a separate joint Registrar may be formed in each state and requisite number of staff may be appointed at different administrative headquarters. Such arrangements will strengthen the working of primaries.

Benefits

(i) Co-operatives aim to participate in the speedy economic development of the country and help in the formation of egalitarian Social System (Jha, 1988).

(ii) The participants will be able to get full price of their products.

Integrated Rural Energy Planning by Rural Energy Co-operatives in Chotanagpur and Santhal Parganas

In Chotanagpur and Santhal Parganas fuelwood, animal dung and agricultural waste constitute 95 per cent of the domestic energy in rural areas. Forest destruction by head-loaders is a common features. Now villagers are realising that fuelwood supplies have to be augumented many folds (Jha, 1988), as the resources in forests are limited in comparison to human as cattle demographic uplift.

In present situation Social Forestry and Biogas technology are the two possible alternatives for solving the rural energy crisis. Beside above, other sources are solar, hydro-electric and wind etc. All the available energy systems shall be combined integrated to solve the energy problems. But in the present situation, large scale energy plantation for fuelwood and smokeless chulha, Biogas technology can solve the problems of tribals womens of Chotanagpur and Santhal Paraganas. In the rural areas burning of fuelwood envelops the indoor environment with heavy smoke and women who have to do all the cooking may be daily exposed to more pollutants, total suspended particulates (TSP) and benzo (a) pyrene (Bap) than even industrial workers. The results are shocking. These womens are suffering from many kinds of smoke related diseases.

In few villages of Chotanagpur Plateau Govt. agencies, volunteer agencies (e.g. Adithi), agricultural University and others have installed smokeless chulha, Biogas plant and solar cooker etc. to mitigate the suffering of women as well as to save fuel. This programme need to cover the entire plateau.

Like other co-operative societies *Rural energy co-operatives* may be established to solve the energy problems of rural areas of Chotanagpur plateau.

Under the rural energy co-operative, tribal people and other weaker section of society shall manage energy plantation, smokeless chulha and biogas systems etc. by making energy as the focal point for the overall economic development and health improvement of women of Chotanagpur and Santhal Parganas.

REFERENCES

Bagchi, S.K. (1986). Fuelwood species. An approach for genetic improvement. *My forest* Vol. 2(2), pp. 107-111.

Jha, L.K. and Jha, K.N. (1985). Social Forestry programme in ensuring better quality of life for the poor. *Myforest*. Vol. 21(4), pp. 317-323.

Jha, L.K. (1988). Forestry schemes to uplift the farmers of Chotanagpur plateau Tec. Bull., B.A.U. Pub., pp. 1-26.

Jha, L.K., Srivastava, K.K. and Jha, G. (1989). Diagnostic survey to study awareness of farmers and existing agroforestry practices in Chotanagpur plateau of Bihar. *Mendal* (in press).

Prasad, V.N. and Jha, L.K. (1985). Creating employment through Social Foestry. *Yojna*, 29(19), p. 11.

Rajan, B.K.C. (1986). Apiculture and farm forestry in semiarid and tracts of Karnataka. *My forest*, Vol. 1, pp. 41-49.

Srivastava, J.N. (1985). Tribal man! Why thy fate is not prosperity. B. Press, Ranchi. pp. 2-3.

Singh, K.P. (1984). Social Forestry Training Programme B.A.U., publication, pp. 20-25.

Chapter 21

Impact of Farm/Agroforestry on Village Socio-Economic Life (A Case Study)

L.K. Jha and Renu Ranjan***

Introduction

The huge population in India brings about resource constrain. The land area of the nation being fixed, rapidly increased population coupled with the developmental schemes has necessitated diversion of forest land for non forest uses to a large extent.

The rural people including tribals living in forest areas are dependent upon the forests in several ways. Tribal life and the forest are so closely related that tribals can't visualise life without forests. The twenty-point programme (1986) has delineated a large number of activities in support of struggle against poverty. However, the mutually supportive package of self employment (IRDP, TRYSEM, DWERA) Wages - employment (NREP, RLEGP) and (JRY) and, area development under DPAR/DDP constitute the areas of poverty alleviation strategy.

The National Commission on Agriculture, 1976 report stress the Socio-Economic importance of social forestry and agroforestry to the rural community as well as in the management of forest resources. The social forestry is a labour intensive scheme and has a huge employment potential. The NCA (1976) has classified Social Forestry into three broad categories, viz., urban forestry, rural forestry and farm forestry (agroforestry). It will thus be seen that the objectives of social forestry are distinctly different from the traditional forestry.

In India, per capita energy consumption in the urban areas continues to be more than ten times of that of the rural areas, even though nearly 74

**Deptt. of Social Forestry, B A.U. Ranchi.*

*** Deptt. of Sociology, Magadh Mahila Womens College, Gandhi Maidan, Patna.*

per cent of the villages have been electrified, only less than 15 per cent of the rural house holds are using electricity (Baidya, 1984), and about 200 to 300 mandays per year are spent by a typical rural family in collecting firewood (Negi, 1986). The villagers are fully acquainted with the importance of cowdung and agricultural residue as manures but they are diverting over 458 million tonnes of wet dung annually resulting in loss of 45 million tonnes of food valued at about Rs. 90,000 million (Jha and Sen, 1991).

On the basis of the report of the Advisory Board on energy present requirement of non-commercial energy is 229 MTCR and it is estimated that this requirement will go upto 280 MTCR by 2000 AD.

Situation in Bihar

Bihar is situated in North East India. The state has total geographical area of 1,73,86,700 ha with two distinct physiographical regions, namely, the Gangetic plains in the northern half and Chotanagpur plateau in the South. The forests in the state extends over an area 23,30,000 ha, mostly in Chotanagpur plateau and account for 16.8 per cent of the total geographical area.

The plateau region of Chotanagpur and Santhal parganas

This region forms part of Zone N. 7 of Agroclimatic zone which is also popularly known as "Eastern Hill region" of the country. Economically the people of this region are more deprived than in other part of this backward state. The tribal population is about 59 lakhs which constitute 13 per cent of the country's tribal population and 8.75 per cent of the total population of the state and 40 per cent of the total population of the region.

The percentage of women workers in total work force during 1971 in tribal Bihar was 14.10 which is expected to rise to 28.89 in 2000 A.D. The higher percentage of women in tribal Bihar is on account of involvement of tribal population in agriculture where women are more active than men.

Bihar social forestry project

The initiation of a social forestry programme during 1978-79 is an important landmark in forestry development in Bihar. This programme has become very popular, and a situation has come when the Government, Semi Government, Universities and voluntary Agencies have joined hands under one umbrella.

As such it was thought worthwhile to evaluate impact of farm forestry/agroforestry programme in socio-economy of villagers of South Bihar. This study will generate knowledge about the success of farm forests agroforestry programme in Chotanagpur plateau of Bihar.

Objectives and Methodology

The present study has been designed with the following objectives:

1. To study the socio-economic characteristics of farmers.
2. To study the traditional and present trend of social/agro/farm forestry.
3. To study the impact of farm/agroforestry in solving the requirements of food, fuel, fodder and timber needs of villagers.
4. To study the total income of the farmers of the villagers.

Methodology

Methodology of this study was devised in such a way that one obtains clear picture regarding living condition of the rural population, their needs, land utilisation pattern, extent of dependence on the forest resources and finally their attitude towards social/farm and agroforestry programme. These information assisted in study on impact of farm/ agroforestry in socio-economics of villages.

The present study was conducted in Ranchi District of Chotanagpur plateau of Bihar. The study was confined to 5 villages of Ranchi Districts; 60 respondent from each village were selected following random selection technique.

Profile of village in brief

(A) *Rarha*: Rarha is a tribal village situated in the hilly areas of Kanke block. The village is surrounded by forests. This village consists of 17 tolas, with total Geographical area 1440 acres. The total population is 1920 of which 92.4% are tribals. The average land holding per household is 4.9 acres.

(B) *Kumharia*: The village is also under Kanke block and is small in size, with 155 families and total population is 1157.70 per cent population are of tribals. The villages are scattered in three pockets. Average land holding per household is 3.2 acres.

Table 21.1 : ***Distribution of Respondents on the basis of Socio-Personal Characteristics (n-60)***

Sl.No.	Variables	Classification	Villages									
			Rarha		Kumharia		Boreya		Sosai		Belangi	
			No.	%	No.	%	No.	%	No.	%	No.	%
1.	Age (Years)	Young (upto 30 years)	3	5.0	27	45.00	19	31.7	11	18.3	14	23.3
		Middle (31-50 years)	35	58.3	32	53.33	39	65.0	35	58.3	43	71.7
		Old (51 and above)	22	36.7	1	0.67	2	3.3	14	23.4	3	5.0
2.	Education	Illiterate	44	73.3	50	83.3	10	16.7	19	31.7	15	25.0
		Read only	3	5.0	3	1.5	6	10.0	8	13.3	8	13.3
		Read & Write only	4	6.7	-	-	12	20.0	6	10.0	8	13.3
		Primary	4	6.7	4	2.5	8	13.3	10	16.7	10	16.7
		Middle	3	5.0	-	-	5	8.3	8	13.3	10	16.7
		High School	2	3.3	2	1.0	15	25.0	6	10.0	8	13.3
		Graduate & above	-	-	1	0.60	4	6.7	3	5.0	1	1.7
3.	Type of family	Nuclear	35	58.3	41	68.3	21	35.0	41	68.3	34	56.7
		Joint	25	41.7	19	31.7	39	65.0	19	31.7	26	43.3
4.	Size of family	Small (upto 5 members)	12	20.0	11	18.3	14	23.3	25	41.7	13	21.7
		Medium (6 to 9)	41	68.3	34	56.7	32	53.3	30	50.0	37	61.7
		Large (above 9)	7.	11.7	15	25.0	14	23.3	5	08.3	10	16.6

(C) *Sosai*: Sosai is a tribal dominated village, located in Mander block with a geographical area of 1326.03 acres and population 1385.59 per cent population are of tribals. Average land holding is 7.8 acres.

(D) *Belangi*: Belangi is also a tribal dominated village (95.9%) situated in the plane area of Ratu block. The total geographical area of the village is around 415 acres and population 532. The average family size and land holding per family is 7.3 and 3.5 acres respectively.

(E) *Boreya*

Socio-Personal Characteristic of the Farmers

Data related to socio-personal characteristics of the farmers are presented in Table 21.1.

The date in Table 21.1 indicate that majority of the farmers in the surveyed villages belongs to middle age group i.e. 31 to 50 years. It is apparent that education brings change in the knowledge, skill, attitude and action of the target group. Data reveals that among the selected village illiteracy was recorded maximum in Kumharia (83.3%), followed by Rarha (73.3%), Sosai (31.7%), Belangi (25.0%) and minimum in Boreya (16.7%).

Table 21.2: Mean value along with their standard errors for personal characteristics of Farmers

Sl. No.	*Variables*	*Rarha*	*Kumharia*	*Boreya*	*Sosai*	*Belangi*
		N=60	N=60	N=60	N=60	N=60
1.	Age	32.83±1.39	31.79±1.23	34.83±1.25	40.53±1.62	43.94±1.72
2.	Education	0.75±0.18	0.50±0.18	2.88±0.25	2.17±0.25	2.33±0.24
3.	Type of family	1.42±0.06	1.32±0.02	1.65±0.03	1.32±0.06	1.43±0.06
4.	Size of Family	6.28±0.21	7.34±0.27	7.43±0.37	6.95±0.38	7.47±0.63

The higher percentage of educated in Boreya and Belangi is due to good educational facilities available there. Besides this, economic conditions also played an important role in increasing literacy rate. In Rarha and Belangi low literacy is due to economic constraints. Their children keep themselves engaged in traditional occupations like cattle grazing etc., because these activities help poor farmers immediately in maintenance of their family.

On critical examination, it is apparent that the rural development depends upon the literacy rate of the villagers. In most of the villages of Chotanagpur and Santhal Parganas more emphasis should be given to increase the literacy level.

The average family size of the household in the village is 5 to 7 persons. However, it fluctuates greatly when considered against land ownership categories. We find that the family size, among landless and medium size categories, is approximately in persons per household. Whereas it increases per household in the marginal and small farmer categories respectively. Data also indicate that majority of the farmers of the selected village belongs to nuclear family system (Rarha 58.3%), Kumharia (68.3%), Boreya (35.0%), Sosai (68.3%), Belangi (56.7%). Whereas in Boreya village 65.0 per cent of farmers living in joint family Pandey (1989) reported that the reasons behind this might be due to christianity influence and urban awareness of the economic necessities which rather influenced the tribals and other weaker sections to restrict the family size.

Occupation

It is evident that agriculture is the main source of income in all villages under survey (Rarha 90.0%), Kumharia and Boreya 88.3 per cent, Sosai 83.3 per cent and Belangi 80.0 per cent.

It has been observed that a chain of head-loaders consisting mostly of women going outside the village with the headload of woods and this explains the prime occupation of those who could not disclose their prime occupation. Since agricultural labour alone cannot provide employment throughout the year, most of the adult population try to seek a secondary occupation to supplement their income from prime occupation.

In general income rate of the respondents is very low. To these persons, any request or appeal not to cut the forest would be nothing but a jock unless alternate source of livelihood is offered.

Table 21.3 : Distribution of respondents on the basis of economic characteristics (n=60)

Sl. No.	Variables	Classification	Villages									
			Rarha		Kumharia		Boreya		Sosai		Belangi	
			No.	%	No.	%	No.	%	No.	%	No.	%
1.	Occupation	Primary (2)	54	90.0	53	88.33	53	88.3	50	83.3	48	80.0
		Secondary (1)	6	10.0	7	11.66	7	11.7	10	16.7	12	20.0
2.	Land holding	Landless (0)	7	11.7	8	13.33	7	11.7	11	18.3	14	23.3
		Marginal (0.1-2.5 acres)	27	45.0	30	50.00	36	60.0	32	53.3	28	46.7
		Small (2.6-5 acres)	18	30.0	15	25.00	9	15.0	12	20.0	12	20.0
		Medium (5.1-10 acres)	6	10.0	5	8.33	4	6.7	4	6.7	5	8.3
		Large (10.1 acres and above)	2	3.3	3	5.00	4	6.6	1	1.7	1	1.7
3.	Socio-economic status (Trivedi sxale modified)	Low (up to 35 points)	55	91.7	39	65.00	36	60.0	44	73.3	35	58.3
		Medium (36-65 points)	5	8.3	19	31.66	18	30.0	13	21.7	24	40.0
		High (66 and above points)	-	-	2	0.02	6	10.0	3	5.0	1	1.7

Note: (1) Occupation :- Farmers having agriculture as primary occupation were assigned 2 score. And other than agriculture were in classification as secondary occupation and assigned 1 score.

Land Holding and Socio-Economic Status

Tables 21.3 and 21.4 pertaining to the land holding indicates that the average land holding of the farmers ranges between 2.79 to 3.75 acres in all the villages. The maximum number of landless farmers was recorded in Belangi (23.3%, Sosai 18.3%, Boreya 11.7%). The majority of the farmers belong to marginal category (Boreya 60.0%, Sosai 53.0%, Kumharia 50.0%, Belangi 46.7% and Rarha 45.0%). The respondents of Boreya had better socio-economic status, more interaction and received more facilities from the other agencies on account of easy acceptability and socio-political consciousness of the inhabitants.

Table 21.4 : Mean value alongwith their standard errors from economic characteristics of farmers

Sl. No.	Variables	Villages				
		Rarha n=60	Kumharia n=60	Boreya n=60	Sosai n=60	Belangi n=60
1.	Occupation	1.90±0.04	1.88±0.04	1.88±0.04	1.83±0.04	1.80±0.05
2.	Land Holding	2.98±0.76	2.79±0.41	3.75±0.59	3.13±0.37	2.99±0.46
3.	Socio-Economic Status	25.76±1.37	29.51±1.94	43.72±2.79	37.01±2.53	36.15±1.97

Note : Calculations based on scores.

Majority of the farmers in all selected villages fell into low socio-economic status group (Rarha 91.7%, Kumharia 65.0%, Boreya 60.0%, Sosai 73.3% and Belangi 58.3%). Highest socio-economic status was observed in Boreya 43.7 per cent and lowest in Rarha 25.76 per cent. The reason for higher socio-economic status in Boreya may be due to larger land holding, good irrigation facilities and the proximity of the village to the Birsa Agricultural University. But now due to various developmental projects started by Birsa Agril. University, Ranchi and other agencies their economic status is improving progressively.

Live Stock and Poultry

In the study report of the villages for appraisal of the proposed, Bihar "Social Forestry Project", it has been mentioned that the study of livestock position indicates that bullocks and Cows are primarily used as drought animals. Whatever, little income generating activities the people are having at present, are all centred around agriculture and other land based persuits in which draught animals are the essentially required. Even the cows are used more as draught animals than for milking purposes. Rearing of goats (as they do) does not involve any additional labour or cost. Likewise poultry or pigs are reared as supplementary sources of their livelihood.

It is apparent from Table 21.5 that in selected villages 73.0 per cent of the respondents have cows, 17.6 per cent, 48.0 per cent, 68.6 per cent, 26.3 per cent, 35.6 per cent and 56.3 per cent of the respondents owned buffalo, bullock, calves, pig, goat and poultry respectively. On an average possession of cattle is maximum in Kumharia, Boreya and minimum in Rarha village. Maximum number of cattle may be due to the fact that some of the respondents had dairy farm with maximum number of improved breed of cows.

Respondents of selected villages are taking interest in the livestock and poultry rearing because they are getting technical know-how from the Extension Directorate of Birsa Agril. University, Ranchi.

Source of Income of Respondents

Rural development is normally the agricultural development but rural transformation as a whole includes development of all the facets of human resources like social, economic, cultural, spiritual and the like of the rural system. Therefore, rural development has to be carried out in its totality rather than as fragmented approaches (Mishra & Srivastva, 1987).

Table 21.6 indicates that agriculture (41.2%) and animal husbandry (29.6%) constitute major share in the income of the total respondents, followed by service (15.8%), forestry (1.2%) and others (3.2%), Agricultural pursuits constituted 45.0% of the total income of the farmers of Kumharia, which is maximum and lowest was at Boreya (39.0%). Maximum contribution of livestock and poultry was observed in Belangi (38.0%) and Boreya (34.0%). Forestry sector contributes maximum income in Rarha (20.0%). Maximum earning (24.8%) from services, wages and labour was observed in Boreya.

Table 21.5 : Distribution of Respondents having Livestock and Poultry

Sl.No.	Livestock and Poultry	Villages											
		Rarha n=60		Kumharia n=60		Boreya n=60		Sosai n=60		Belangi n=60		Overall n=300	
		No.	%	No.	%	No.	%	No.	%	No.	%	No	%
1.	Cow	28	46.7	50	83.3	50	83.3	46	76.7	45	75.0	219	73.0
2.	Buffalo	3	5.0	18	30.0	15	25.0	10	16.7	7	11.7	53	17.6
3.	Bullock	19	31.7	22	36.7	36	60.0	30	50.0	37	61.7	144	48.0
4.	Calves	18	30.0	53	88.3	51	85.0	41	68.3	43	71.7	206	68.6
5.	Pig	20	33.3	16	26.7	14	23.3	10	16.7	19	31.7	79	26.3
6.	Goat	21	35.0	20	33.3	21	35.0	20	35.0	25	41.7	107	35.6
7.	Poultry	28	46.7	40	66.7	30	50.0	35	58.3	30	50.0	169	56.3

Table 21.6: Percentage contribution of different sectors in total income of respondents (approximately)

Sl.No.	Items	Villages					
		Rarha	Kumharia	Boreya	Sosai	Belangi	Pooled
		n=60	n=60	n=60	n=60	n=60	n=300
		%	%	%	%	%	%
1.	Agriculture	40	45	39	42	40	41.2
2.	Livestock/Poultry	25	26	34	25	38	29.6
3.	Forestry	20	15	1	10	5	1.2
4.	Service (including Wages/labour)	10	11	24	19	15	15.8
5.	Others	5	3	2	4	2	3.2

The main reason of increasing economic status of farmers is successful implementation of developmental programmes by the Govt. agencies as well as Birsa Agril. University, Ranchi. In the beginning of the programme a lot of difficulties was encountered in creating an atmosphere conducive to win the goodwill and confidence of the villagers. Many field camps were organised to arouse their consciousness and awakening towards the advantages to be had from the adoption of improved technology. With the introduction of improved agriculture technology, the agricultural productivity in terms of acreage yield has gone up in these villages. Similarly, adoption of improved animal husbandry and forestry practices the additional income generation of farmers has gone up. The above view is also supported by Mishra and Srivastra (1987).

Fodder, Foliage, Fuel and Timber Requirements of the Villages

Fodder and foliage

Actually, in most of the villages, there is no organised system of fodder collection activity. The entire system is primitive. The livestock are left generally in the barren land, in the nearby forest area. Farmers are not selective regarding type of fodder. The total requirement of fodder and tree folliage per year is recorded 450.04, 799.36, 709.18, 630.72, 601.52 tonnes in Rarha, Kumharia, Boreya, Sosai and Belangi respectively. In general villagers face the problem of collection of fodder, because there is no grazing pasture land in the village and fodder cultivation is not practised.

In Kumharia village farmers are self sufficient. They grow grasses alongwith tree species in the wastelands. In 17 acres land farmers harvested 13.49, 16.49, and 18.49 quintal sundried grass in the year 1987, 1988 and 1989 per acre. In another 50 acres of wastelands farmers did not grow even a blade of grass. But under silvipastoral system, farmers harvested approximately 517.6 quintal of fodder in 1988 and 697 quintal in 1989. In Rarha and Belangi village farmers also endeavoured to grow fodder in their wastelands under the supervision of the experts of B.A.U., but they did not harvesting sufficient quantity of fodder.

Fuel

Table 21.8 presents total annual firewood requirements which is 788.40, 985.20, 1095.00, 985.20, 1095.00 quintal in Rarha, Kumharia, Boreya, Sosai and Belangi respectively when total annual requirement of fuel was calculated in sampled villages, it came around 1477.80, 1832.40, 1576.20 and 1642.20 quintals in Rarha, Kumharia, Boreya, Sosai and Belangi respectively.

Table 21.7 : Average and total requirement of fodder and foliage per family per day

Sl.No.	Villages	Total No. of large and small cuminants	Tree Foliage			Fodder		
			Average require-ments kg/ day per animal	Total require-ment of Tree Fol-iage per day in kg.	Total require-ment of Tree foliage per year (in Tons)	Average require-ments kg/day per an-imal	Total re-quire-ments of per day/ kg.	Total require-ments of Fodder per year (in Tons)
1.	Rarha (N=60)	137	4	548	200.02	5	685	250.02
2.	Kumharia (N=60)	219	3	657	239.82	7	1533	559.54
3.	Boreya (N=60)	217	2	424	154.76	7	1519	554.53
4.	Sosai (N=60)	192	3	576	210.24	6	1152	420.48
5.	Belangi (N=60)	206	3	618	225.57	5	1030	375.95

Total requirement of fodder and Tree foliage per year (in ton)

	Rarha (N=60)	Kumharia (N=60)	Boreya (N=60)	Sosai (N=60)	Belangi (N=60)
Total requirement of Fodder and Tree Foliage per year (In Ton)	450.04	799.36	709.18	630.72	601.52

From the response of respondents it is apparent that the local community uses wood as a major source of fuel both for cooking and other purposes; since there is practically no body to prevent them collecting fuelwood from forest they appear to be lavish in fuelwood consumption. Moreover, it is very difficult to get the accurate information about the quantum fuelwood consumed for domestic needs.

The problems of collection of fuelwood revolves round the distance, hazard in the exercise of cutting in the wood, stocking it, preparing the bundle and then carrying it to the village.

It is expected that after few year most of the respondents will be self sufficient in collecting fuelwood from their own farmland due to agroforestry practices adopted. In Kumharia village in the year 1988 about 500 trees were harvested by the farmers from their own farm to fulfil the domestic requirement of fuelwood. Currently about 250 acres land is under tree cover. In next few years, they will be self sufficient with regard to domestic fuelwood requirement. In Boreya and Belangi villages Govt. agencies introduced Gobar gas plant etc. Thus, their dependence on forests for fuelwood is decreasing day by day (Jha & Kumar, 1990). Similarly, in other villages this programme should be started after motivating villagers to protect and manage the plantation.

Timber

The total requirements of the timber poles of the respondents come around 300.0, 360.0, 480.0, 420.0 in Rarha, Kumharia, Boreya, Sosai and Belangi respectively. In the sample villages most of the roofs of the houses are constructed with small size poles. The roof requires frequent repairing. As per estimation, on an average 4 to 5 cft timber is utilised (Jha & Singh, 1990).

To solve this problem, farmers started tree plantation in their land in most of the selected villages. The most successful work in this connection was done by farmers of Kumharia and Belangi villages. In near future they will be self of sufficient with regards to their requirements of small timbers.

Extent of Awareness Regarding Social/Agro and Farm Forestry

Result presented in Table 21.9 reveals that maximum 71.7 per cent of farmers of Kumharia adopt single row bund plantation, whereas only 28.3 per cent farmers of Sosai village do this. In case of double row bund plantation, maximum percentage was recorded in Kumharia (51.7%).

Table 21.8 : Fuel requirements of the villages

Sl. No.	Variable	Average requirement per family per day N=60 (in kg)					Average requirement per family per year N=60 (in quintals)					Annual fuel requirement of villages N=60 (in quintals)				
		1	2	3	4	5	1	2	3	4	5	1	2	3	4	5
1.	Fire Wood	3.60	4.50	5.00	4.50	5.00	13.14	16.42	18.25	16.42	18.25	788.40	985.20	1095.00	985.20	1095.00
2.	Agricultural Waste	2.00	1.00	1.25	1.50	1.50	7.30	3.65	4.56	5.47	5.47	438.00	219.00	219.00	228.20	328.20
3.	Cow dung Cake	1.15	1.50	0.75	1.20	1.00	4.19	5.47	2.73	4.38	3.65	251.40	328.20	163.80	262.80	219.00
4.	Total of all kinds of fuel in villages	6.75	7.00	7.00	7.20	7.50	24.63	25.54	25.54	26.27	27.37	1477.80	1532.40	1532.4	1576.20	1642.20

Table 21.9 : Extent of awareness regarding social forestry/Agroforestry practices

Sl. No.	Statement	Respondent	Villages											
			Rarha n=60		Kumharia n=60		Boreya n=60		Sosai n=60		Belangi n=60		Pooled n=300	
			N	%	N	%	N	%	N	%	N	%	N	%
1	2	3	4		5		6		7		8		9	
1.	Do you know about single row bund plantation	A	36	60.0	43	71.7	28	40.7	17	28.3	42	70.0	166	55.5
		NA	24	40.0	17	28.3	32	55.3	43	71.7	18	30.0	134	44.6
2.	Do you know double row bund plantation	A	16	26.7	31	51.7	16	26.6	12	20.0	27	45.0	102	34.0
		NA	44	73.3	29	48.3	44	73.4	48	80.0	33	55.0	198	66.0
3.	Do you know block plantation of timber species	A	37	61.7	31	51.7	15	25.0	18	30.0	16	26.6	117	39.0
		NA	23	38.3	29	48.3	45	75.0	42	70.0	44	73.4	183	01.0
4.	Do you know block plantation of silk and Lac hos plants	A	11	18.3	12	20.0	18	30.0	13	21.6	15	25.0	69	23.0
		NA	49	81.7	48	80.0	42	70.0	47	78.4	45	75.0	231	77.0

Table 21.9 — Contd.

1	2	3	4		5		6		7		8		9	
5.	Do you know block plantation of fodder trees	A	5	8.4	12	20.0	8	13.3	2	3.3	5	8.4	32	10.7
		NA	55	91.6	48	80.0	52	86.7	58	96.70	55	91.6	268	89.3
6.	Do you know plantation of timber species alongwith superior quality of grass	A	-	-	15	25.0	8	13.3	4	6.6	9	15.0	36	12.0
		NA	60	100.00	45	75.0	52	86.7	56	93.4	51	85.0	264	88.0
7.	Do you know plantation of timber/ fruit species along with natural grass	A	24	40.0	38	63.4	13	21.6	15	25.0	21	35.0	111	37.0
		NA	36	60.0	22	36.6	47	78.4	45	75.0	39	65.0	189	63.0
8.	Do you know plantation of fodder trees alongwith grass	A	4	6.7	11	18.4	5	8.4	-	-	4	6.6	24	8.0
		NA	56	93.3	49	81.6	55	91.6	60	100	56	93.4	276	92.0
9.	Do you know about silvi-pastoral practices three tire system	A	-	-	4	6.6	-	-	-	-	2	3.4	6	2.0
		NA	60	100	56	93.4	60	100	60	100	58	96.6	294	98.0

Table 21.9 — Contd.

1	2	3	4		5		6		7		8		9	
10.	Do you know growing agricultural crop alongwith tree species (scientifically)	A	-	-	8	13.4							8	2.7
		NA	60	100	52	86.6	60	100	60	100	60	100	292	97.3
11.	Do you know growing Agricultural crop alongwith tree species traditionally	A	55	91.7	57	95.0	48	80.0	47	78.4	51	85.0	258	86.0
		NA	5	8.3	3	5.0	12	20.0	13	21.6	9	15.0	42	14.0

A = Aware NA = Not Aware

bout the block plantation m imum percentage of awareness was obser ed in Rar 1.7%) minimum in Boreya (5.0% .

ery poor response was observed in case of block plantation of odder trees, only 10.7 per cent of total respondents was not aware of plantatio of tim er s ies along wit superior uality of grass. hen respondents were asked regardin growing o natural grass ong with 'mber fruit tree species, 3. per cent espondents of Kumharia e hi ited there knowledge and awareness.

In case of traditional agroforestry 5 per cent of umharia 1. per cent of Rarha and 80.0 per cent Boreya, 78. per cent of osai an 85.0 per cent of Belangi s owed their awareness, 7.3 per cent of the pooled respondents were not aware of scientific develo m nt in the field of farm agroforestry.

It is apparent that in general farmers showed awar ness regarding social fores y/agroforestry prog m s m imum a areness being observed in Kumharia villa e. They were ac uainted with most of the component of e soci forestry pr grammes.

The response as satisfactory i oreya and, R ha, Belangi farmers howe er did not h ve knowledge egarding improved scientific farm agroforestry practices.

hus unde present circ ms ces i is the duty of th exper s xtension wor ers and others to tra s er latest t hno ogy related to agroforestr practices to the field. Bes'de this scientists/foresters shoul cond ct ex 'ments to o the sui ble ees cies, suitable co b'nation of trees and grass f r each agrocl'matic one.

t nt f opt on a mpact f ocia fo t in illa s

arge armers adopted more soc' /a roforestry programme and the l dless showed e least interest. The extent of doption in case of marginal, small d the medium farmer , almost e s e. sponde a opt d more social/agroforestry progr me on astelands. mer saw more benefit on hese than agricultural land. Tables 10( ) to 1 10( ) dealing with the adop 'on of lan tion reveal that umharia farmers adopted bui d l tation (8.33%) plantation of fruit tr s (0 o), lock lantation of misc species (0%), silvi pastoral system (50 0 o) and agrisilviculture 38.3%.

Table 21.10 : (A) Extent of adoptation of social forestry/Agroforestry practices in the villages

Practice Items	Bund Plantation									
	Rarha n=60		Kumharia n=60		Boreya n=60		Sosai n=60		Belangi n=60	
	No.	%	No.	%	No.	%	No.	%	No.	%
Bund Plantation:										
(i) Single row timber species plantation	-		5	8.33	10	16.67	5	8.33	5	8.33
(ii) Double row timber species plantation	-	-	-	-						
(iii) Plantation of fruit trees on bund	1	1.67	-	-	5	8.33	2	3.33	6	10.00
(iv) Plantation of fodder trees on bund	-	-	2	3.33	-	-	-	-	2	3.33
(v) Plantation to solve the problem of fuelwood			4	6.67	4	65.67	3	5.00		
	1	1.67	11	18.33	19	31.67	10	16.67	13	21.67

Table 21.10(B): Block plantation adoption

Practice Items	Rarha n=60		Kumharia n=60		Boreya n=60		Sosai n=60		Belangi n=60	
	No.	%	No.	%	No.	%	No.	%	No.	%
Block Plantation										
(i) Block Plantation timber species	4	6.67	10	33.33	6	10.00	1	1.67	8	13.33
(ii) Block Plantation for fuelwood and small timber	2	3.33	15	25.00	-	-	4	6.67	5	8.33
(iii) Block Plantation of arjun and Mulberry for Tassar silk worm rearing respectively	-	-	5	8.33	1	1.67	2	3.33		
(iv) Block Plantation of Lac host plants										
(v) Block Plantation of fodder trees	3	5.00	2	3.33						
Total :	9	15	42	70	7	11.67	7	11.67	12	20

Table 21.10(C): Silvipastoral system practices

Practice Items	Rarha n=60		Kumharia n=60		Boreya n=60		Sosai n=60		Belangi n=60	
	No.	%	No.	%	No.	%	No.	%	No.	%
Silvipastoral of timber										
(i) Plantation System species alongwith superior quality of Grass.										
(ii) Plantation of timber species along with Natural Grass	15	25	30	50			2	3.33	8	13.33
(iii) Plantation of fodder trees alongwith grass species			5	8.33						
(iv) Silvipastoral practices in three tier system										
(v) Plantation of timber tree species, fodder species and good quality of grass										
	15	25	35	58.33	-	-	2	3.33	8	13.33

Table 21.10(D): Agri-silviculture practices

Practice Items	Rarha n=60		Kumharia n=60		Boreya n=60		Sosai n=60		Belangi n=60	
	No.	%	No.	%	No.	%	No.	%	No.	%
Agri-silviculture practices										
(i) Growing of Agricultural crops alongwith tree species as suggested by the extension worker	2	3.3	21	35.0	-	-	4	6.6	-	8.3
(ii) Growing of Agricultural crop alongwith tree species traditionally	3	5.0	2	3.3	5	8.3	-	-	3	5.0
Total	5	8.3	23	38.3	5	8.3	4	6.6	8	13.3

In Kumharia, extent of adoption was the maximum in comparison to other villages.

The extent of adoption of social forestry/farm and agroforestry largely depends on appropriate technological inputs. This is particularly very relevant in rainfed farming.

Conclusion

Adoption of above programme helped in creation of employment opportunities in the village.

The long term physical assets generated in village in term of tree belt can't be easily quantified in terms of quick ready income, but this will continue to pay heavy dividends to the villagers for even in future in the form of ecological security.

The above mentioned target can be achieved by rapid progress in the field of social forestry/farm/agroforestry. The rapid progress in plateau areas can be a reality only when there is mass involvement of tribals in it. They possess capacity to participate actively in the various activities of social forestry programme.

Tribals may plant trees in their marginal and submarginal lands. They may be encouraged by the Govt. to plant trees in the degraded forest areas lying near their villages under participation management system. They may adopt farm forestry in their lands to meet the rapidly growing requirements of fuel, fruits, fodder, food grains etc. It becomes the duty of extension staff to examine the social conditions of farmers and motivate them to adopt farm/agroforestry so that their income may rise rapidly. The extent of participation of farmers in farm/agroforestry depends on the nature and extent of motivation by the extension staff.

Thus to make this programme successful in real sense at large scale, we have to provide inducement for the involvement of the mass and concerted efforts are to be made to provide various incentives to make this programme a grand success. Several efforts have been made by the Govt. on this score, but the pace of work cannot be considered up to the desired level as yet, we may expect that concerted efforts will be taken to provide all possible incentives to the participants.

Chapter 22

Panchayat Raj Institutions and Social Forestry/Agroforestry Schemes : How to carry out the Schemes

*S. Parameswarappa**

With the establishment of the Zilla Parishads and Mandal Panchayats in 1987, Social Forestry Schemes like many other development schemes are being implemented by these panchayat raj institutions.

The work 'implemented' is an omnibus word. It can mean anything from just approving the programme to actually executing it out through agencies other than the departments concerned. It can also mean routing money through these institutions and carrying out the works through the concerned departments.

Social Forestry including agroforestry is different from traditional forestry as it has a large number of social features. Some of the salient social features in social forestry are :

(i) The programme is implemented on lands belonging to the people, i.e., the village common lands/open access lands and the private lands. These common lands are the gomal lands, C and D class of lands, tank foreshores, roadsides and canal sides.

(ii) These common lands are under utilised at present. The social Forestry programmes envisage optimum utilisation of common lands and the under utilised private lands and the marginal agricultural lands through soil and water conservation and afforestation activities.

(iii) With the afforestation of these under utilised lands, it is envisaged to produce fuelwood, small timber, bamboo, fodder, green

**Principal Chief Conservator of Forests, Karnataka, Bangalore.*

manure, usufructs and other forestry products and to give these outputs to the rural communities free of cost or at subsidised rates and thereby mitigate their hardships which they are currently put to in procuring these forestry products.

(iv) Sustained employment generations is an important direct benefit from these Social Forestry which is labour intensive. The programme can be implemented in every nook and corner of the country and can be carried out all the year round.

(v) Improvement of the micro climate due to soil and water conservation and afforestation is another important aspect of social forestry activities.

Social forestry/agroforestry programmes are not simple afforestation programmes. A series of processes is involved to activate the social content. Which are as follows :

(i) Consultation with the local people to :

(a) Enumerate the extent of land resource available for afforestation;

(b) determine the needs of the people;

(c) Find out the interest the people have in these programmes and plan ways and means to activate interest.

(d) Schedule the time required for phasing the afforestation programme;

(e) determine the choice of species;

(f) determine the outputs and quantity that are likely to occur from the programme; and

(g) Plan out the administrative and silvicultural management of the assets created.

(h) The whole process is termed as micro-planning is carried out at the grass root level and is therefore need based. These are :

(ii) Raising the required number of seedlings.

(iii) Advance work of site preparation viz., pitting, trenching/mouding, V ditch (furrow) formation etc.

(iv) Planting the seedlings.

(v) Weeding, application of fertilizers and pesticides, soil working and fire tracing.

(vi) Preparation of management plans for the plantations and thereafter carrying out the prescriptions of the management plans like -

(a) Pruning;
(b) thinning
(c) lopping or pollarding
(d) usufruct collection
(e) final felling;
(f) sharing and distributing the out puts;
(g) replanting;

These operations are envisaged to be spread over a period of 10-30 years and in case of usufruct yielding species to over a hundred years. Annual collection of grass will however go on, once the grass and legume (Hamata) cover gets established in about 2 years.

Now, who is to carry out all these works? Which would be the best agency to carry out these works with continuity? Which would be the best agency that would have sustained interest in carrying out these activities? These are a few issues that need to be addressed.

The Zilla Parishads and the Mandal Panchayats at best can only be the agencies to allocate outlays to the social forestry programmes in their annual plans and to release the money to the implementing agency according to the outlay. They cannot do anything more than this in view of the variety and quantum of developmental works they normally have to undertake every year. In view of the fact that the financial resources are inadequate to meet all development activity and in view of the fact that social forestry activity gets the least priority and often is not given any outlay, it becomes necessary to make it mandatory on the Panchayat Raj Institutions to earmark 25-30 per cent of its annual expenditure on social forestry schemes in addition to ploughing back into afforestation a portion of the revenue which these institutions would get from the older plantations. Although social forestry programmes are important and are labour intensive, since they take a long time to show results, Panchayat Raj Institutions tend to spend money on works that show quick result like road formation, well sinking, building construction etc, to gain public popularity. The members of these institutions after all, have a desire to be relected and naturally they tend to gain popularity by implementing schemes that show quick results. Therefore it must be made mandatory on these institutions to spend 25-30 per cent of their annual budget on social forestry programmes in addition to ploughing back in afforestation a portion of the revenue which these institutions would get from the older plantations.

The best agency under these circumstances to implement social forestry programmes, would be forestry wing of the Zilla Parishad for the time being. A lot of staff has been transferred to the Zilla Parishad and there is a separate establishment at the Zilla Parishad, right from the Deputy Conservation of Forests (Social Forestry) Zilla Parishad, to the grass root level motivators. This establishment of the forest department in the Zilla Parishad is the best agency, for the time being to implement the social forestry programme.

Social forestry/agroforestry programmes are under implementation on a large scale since last seven years. Many of the assets created are due for harvesting. These assets are required to be shared and distributed amongst the people in a fair and equitable manner. At present, the Government order specifies that these plantations raised under Social Forestry are to be handed over to Mandal Panchayat for maintenance and protection and sharing after harvest. In reality, not many panchayats are enthusiastic to take over these plantations as there is no mechanism in Mandal Panchayats to handle this. For this purpose, we envisage setting up of Village Forest Committees in every village. These Committees are envisaged to be made up of 13 to 20 members or a little more if necessary and shall be comprised as follows :

(i) The mandal pradhan/upa-pradhan or member elected from the village as Chairman of this Committee.

(ii) The forester having jurisdiction over the area as member secretary.

(iii) The forest extension worker having jurisdiction over the area as member.

(iv) The village accountant as Member.

(v) 30 per cent of the members of this committee shall be women and these women members shall be from landless, marginal and small farmer households.

(vi) 25 per cent of the members of this committee shall be from the SC/ST and these members shall be from the landless, marginal and small farmer households.

(vii) There shall be one member from the artisan group in this committee.

(viii) The remaining members shall be from other walk of like including one representative of a NGO/Voluntary Organisations if they are working is the region.

The members listed at serial No. (v) to (viii) shall be elected in a Grama Sabha conducted by the mandal panchayat. The tenure of these committees in stated to be for a period of three years from the date of appointment.

The aims and objectives of the village forest committees for the time being is envisaged to be as follows :-

(i) To assist the mandal Panchayats and Zilla Parishad in planning and carrying out the social forestry programmes in their respective village.

(ii) To oversee the management and distribution of the social forestry plantation resources in accordance with the management and distribution plans.

Further, the present Government order of 23-1-1986 on sharing of produce of Social Forestry Plantations may be modified as follows :-

(a) The order should be made applicable to plantations raised on gomal, C & D lands, foreshores, roadsides and canal sides under all social forestry schemes (currently) it is applicable to block plantation raised under World Bank/ODA scheme.

(b) The order should also be made applicable to roadside and canal side plantations raised under World Bank/ODA scheme (currently it is applicable to block plantations only).

(c) The order should make it mandatory on the Zilla Parishad and Mandal Panchayats to plough back for developmental works, all the revenue they receive from these plantations, in the same village from which the plantation revenue was realised. 50 per cent of this revenue must be mandatorily ploughed back into afforestation activity in the same village (the present order does not say as to what the panchayats should do with the plantation revenue),

(d) The order should be modified to make the mandal panchayats sole beneficiary for the produce from these plantations. Half of this produce shall be given by the mandal panchayats, free of cost, to the landless, marginal and small farmer of the same village, as identified by the Village Forest Committees, where plantation was raised. In an event there is surplus, the panchayat may sell it along with their share or entrust the sale to the forest department or to the Village Forest Committee (The present order states that the panchayats would get 50 per cent of the produce, half of which they may give to landless, marginal and small farmers as identified by DRDS at concessional rates and the balance panchayats may sell it or entrust the sale to the forest department).

(e) The sale shall first be done in the village where the produce was grown for the benefit of the local villagers. In an event there is excess, which is to be determined by the Village Forest Committees, it may be sold to other villagers of the mandal and thereafter to people outside the mandal. (Present order does not speak on the manner of sale).

(f) All the fodder, green manure, prunings, lops and tops shall be given to the landless, marginal and small farmers of the same village as identified by the Village Forest Committees, free of cost. (Present order speaks of fodder to be given free of cost).

(g) The fruit bearing trees shall be marked and numbered and 50 per cent of the trees shall be assigned to the landless, marginal and small farmers of the village for collection of usufructs free of cost. The fruits of the remaining 50 per cent of the fruit trees, the panchayat shall dispose in the manner indicated above. (Present order does not speak on fruit trees).

With necessary training to be provided by the forest department and the experience to be gained it is envisaged that these Village Forest Committees will be capable of carrying out the following works in social forestry schemes in addition to (i) to (ii) above :-

(i) Micro-planning

(ii) Extension;

(iii) Afforestation;

(iv) Preparing management plans for harvesting and sharing of the produce.

It is envisaged that these committees will have sustained interest in social forestry programmes since these committees are made up of local people themselves.

We envisage these committees to become capable of doing all the above works in the village themselves, having linkage with the forest department through the member secretary and forest extension worker. These committees would of course require funding for carrying out the afforestation activities. The required funds may be given to them by the Zilla Parishads through the Mandal Panchayats. Initially all the funds required may be provided for first generation plantations but at a later stage for second and later generation plantations, portion of the money required is to be ploughed back by the mandal panchayats from the revenue they would receive from the first generation social forestry plantation. With this transfer of activities to the Panchayat Raj institutions, social forestry programmes could be a people's movement in the true sense of the word in Karnataka.

Chapter 23

Psychological Antecedents and Consequents of Social Change and Development : Implications for Social and Agroforestry

*L.N. Singh**

The term 'development', particularly in the Third World countries, perhaps is the most commonly shared concern; and there is constant anxiety to reach the prosperity, even more than, of the advanced countries of the world. The basic characteristics of the latter is their sustained economic growth through the process of industrialization based on vast investments of capital, high level application of science and technology to productive processes and organizational functioning. It has resulted in a degree of economic development and improvement in material well being of the people. But in the wake, it has implied concomitant changes in the environment and society that has often had deleterious consequences for the very people for whom development has been designed. Viewed in these perspective, the term development in true sense has aroused the feeling where there is exploitation of natural resources, pollution of atmosphere, disturbance of ecological balance, creation of new health hazards, deterioration of social life, accentuations of social and economic inequalities, generation of unhealthy tension, and even tolerance of corruption and other social ills simply because they are looked upon as inevitable to the very process of rapid growth.

At the outset, the level of economic prosperity that many countries of Europe and American continents have achieved became the norm which the underdeveloped countries of the World set about to attain within a shortest possible span of time. Development strategy was based

---

**Deptt. of Psychology North-Eastern Hill University, Mizoram Campus, Aizawl-796007.*

on an uncrucial emulation and extrapolation from the experiences of the economic growth model of western countries grossly disregarding the fundamental differences in socio-cultural constraints and the local conditions and circumstances. For social scientists Max Weber (1958) had set the trend. He ascribed the failure of economic development for Asian countries to contain deeply rooted attitudes included by religion and the absence of protestant work ethic which was regarded as the carnstone of economic revolution of the west. Socio-cultural features of traditional societies were contrasted with those of so-called "modern" societies, the former being often considered an intervening with the functioning of the economic and technological forces and, therefore, required to be changed and modernized (Rostow, 1952; Lewis, 1955). Emphasis on achievement motivation as proceeding economic development (McClelland, 1961), its inculcation among entrepreneurs leading to the hightening of economic activities (McClelland and Winter, 1969), and signal importance attached to attitudinal modernity (Inkeles and Winter, 1974) are typical examples of psychologists approach to development and the assertion of his role in the process. Dicholomizing the attitudinal characteristics of traditional and modern societies, Triandis (1971, 1973) had identified a "syndrome of characters" which was considered vital for all utilization of modern technology. The model implied that development was possible only through the diffusion of modern technology from the west which also required the adoptation of management practices, patterns of interpersonal relationships and attitudes and value systems.

As the development is centered upon the man, his abilities and creative faculties; the socio-cultural factors and values are now well recognized both as the determining factors (antecedents) and ultimate results (consequents) of development when the evaluation of successfulness/unsuccessfulness or adoption/rejection of the development plans are viewed (Berry, 1980; Sinha, 1988; Tiri et al; 1988). This is because, other things being equal, ultimate objectives of development strategies are to meet the true needs and aspirations of peoples in order to ensure their genuine fulfilment However, the development programmes and policies are formulated and implement without considering the human factors in relation to their ecological and socio-cultural perspectives which is now a well recognized precondition for endogeneous development, means the development from within the sources for integrity and development. Thus this becomes the reminder of the article to demonstrating the theoretical conceptualization and associated problems of psychological intecedents and consequents of social and cultural change and development in culture specific and cross-cultural perspectives, and finally, to derive implications for social and agroforestry for attention while formulating the development programmes for upliftment of back trodden class of people. To meet these objectives, the present review article is selectively attempted on the psychological antecedents and consequents of social change and development, and while this selective attention, selective referrals of other predominant approaches of economic, sociological and anthropological are made, however, a full account on these perspectives would be desirable.

Conceptual theorization and associated problems

There are essentially three conceptual problems at the socio-cultural level of analysis regarding social and cultural change and development: (i) the value-laden nature of the concept of development and modernization, (ii) the universality of the processes and stages which mark the various stages of process, and (iii) the linearity of change.

At the socio-cultural level of analysis, the definition of social change with development has not been a new phenomena. Nisbet (1971, pp. 98-99) pointed out that "social change was conceived after the pattern of organic growth", as early as classical times. However, the narrower identification of development with modernization appeared during and following the period of western colonial expansion (Omvedt, 1971, p. 122). The historical and intellectual basis of this identifications has been located in Mazrui (1968, p. 69) in the tradition of "social Darwinism" complete with its overtones of "racism and ethrocentrism" (p. 70). The forms of social and cultural change most often studied are now value-free; of course as a normal outgrowth of values of the western societies in which the social science tradition have been developed. Be it is, none of the assertions suggest that the studies is wrong, however, the value bias is clearly present which forms the basis for social and cultural change at individual level.

A second problem is the universality of the phenomena of social change. This is epitomized with the question "are the part and end in the state similar everywhere all the time?", "are "traditional"-"modern" in the state the poles between which all societies and individual pass?" or is the part are processed similar in all cases. On all these counts, at least three varieties of answers may be descerned from the literature on social and cultural change and development. The first assert that both.process and states are similar everywhere, that certain feature of traditional life are the same, that the modern goals are shared, and that common steps are (or need be) taken to move between the two poles. The second position states that they are only local and unique phenomena, that traditional cultures vary widely, that goals that are sought vary as well (and may indeed not be "modern") and that the steps taken in rote must necessarily vary. And finally the third general position maintains that the processes and the point are similar everywhere, but care must be taken in considering the probable existence of traditional cultural variation at the beginning point. All three positions are maintained by those studying individual behavior in relation to social change, and the question of "universality" of the process perhaps seems to be beyond agreement.

Third, is the process of change essentially linear, or is there "re-affirmation" of tradition following exposure to acculturative stress. Jahoda (1961) reported such re-affirmation of phenomena in West Africa, as much earlier Park (1928) considered wonder resolution to marginality to be "to swing about the reaffirm" one's traditional culture. Gusfield (1967) has drawn up "seven fallacies" in and "linear theory of social change", including assumption about traditional homogeneity, and conflict between "traditional"-"modern" forms of behavior. Furthermore, Kavolis (1970) suggested that "... either that the process of modernization of the personality is in some respect being reversed after modernization has been accomplished or that the conceptualization of psychological modernization is too narrow and has never encompassed all of the essential transformations of the personality during modernization (p. 435). Kavolis further argued that many of the behavior that are included for discussion on social and cultural change are being repudiated by many young people in "modern society": an interest in mysticism, and a lost of faith in science and technology (among many indicators) point to the emergence of a "post modern man". A critical analysis of both "modern" and modernizing society reveal that there is some evidence that suggest that the drift of behavior change is not linear, that traditional or modern values and behaviours are increasing in frequency. For those studying individual behaviour during social change, it appears unwise to ignore the non-linearity of the phenomena. This is the point where more emphasis on individual change and development are being advocated in relecence to the indegeneous strategies for development. And simple observation of disparity in the rate of development, despite uniform implementation of developmental programme, it is now strongly felt that as the different communities reacted in their characteristic ways for adoption of the developmental programmes (without considering the traditional socio-cultural systems and practices, and their genuine expectations and fulfilment of their demands), hence disparity in the rate of development. This is the point on which the present review is attempting to focus within the materials of psychological antecedents and consequents of social change and development.

Within the materials on antecendents to change there are two large traditions, and third set of smaller but significant programmes. These first two are the general area of attitude and beliefs (with workers such as (Inkeles, Dawson and Jahoda), an area of achievement orientation (McClelland, and de Vos); the third set includes many other psychological variables such as authoritarian socialization (Hagen), operant conditioning (Guthrie) and modern personality (Wallae). However, among these many approaches of psychological studies on attitude and beliefs, achievement orientation achievement means and achievement goals - the motivational characteristics in relation to development paradigm and acculturative stress paradigms are included selectively so as to focus on the psychological antecedents and consequence of social change and development.

Behavioural Model of Social Change and Development

There is a basic division in the social change and development literature between those who view change primarily in terms of material gains (factors) or the change and development centered upon man, his abilities and creative faculties, and socio-cultural factors. This division was apparent in the classical volume edited by Weiner (1966), and in literature (Desai, 1971; Kunkel, 1966; Campbell and Converse, 1972) it remains as a basic issue. On these counts almost all the theoretical statements and empirical studies stress emphasis on either the psychological antecedents to change, or the psychological correlates or consequents of change. And given the need to attend to the psycho-social-cultural features of social change and development, the available empirical treaties suggest five major distinctions of psychological importance : the locus, source, direction, dynamics and sequence of social change and development (Berry, 1980).

The first issue (locus of change) pertains to the location of change and its level of analysis. The three different loci : socio-cultural, institutional, and individual find mention in literature. At the socio-cultural locus, social change involves large systems such as nations, regions or the cultural groups and the level of analysis tends to be anthropological, political, economic or micro sociological. At the institutional locus, social change involves economic or governmental institutions, and the level of analysis tended to be in economic or sociological terms. At the individual locus, the change involves attitudinal, motivational and cognitive characteristics, and the level of analysis tends to be socio-psychological (Kelman and Warwick, 1973). As regards the source of change the external and internal sources, although the change process could not be attributed to single factor, are attributed. The former focuses on the cultural diffusion/the developmental programmes, while the later lie in the internal social/psychological dynamics of the cultural groups.

The cultural biases inherent in the terms as traditional and modern comprehend some basic dimensions of change. Three general directions of change appear in the available empirical literature : (i) the direction of becoming "modern" in the usual sense of modernization and homogenization of world cultures; (ii) the direction of a "traditional" life style where there is a reaffirmation of characteristic values; and (iii) some "novel" life style on a dimension that is independent of usual "traditional-modern" axis. Forthly, the dynamics of change with its distinction between the process of change and the state that exist at some time (point) during the process, and finally, the sequence of change and development —the psychological antecendents from psychological consequents - deserves special mention and attention for precise scrutiny and understanding of the impact of development programmes (strategies) on motivational and cognitive characteristics at the individual level - the man - in social and cultural perspectives for whom the programmes are formulated and implemented. The interplay and relative dominance of the theoretical constructs and the processes lead three separate features of schematic overview of the voluminous literature on social and cultural change and development : (i) the structure, (ii) the relationships; and (iii) the content (Berry, 1980).

With regard to the structure, the overview is arranged into two levels (the socio-cultural and individual, with the institutional subsumed under the former); it is arranged horizontally by sequence (with antecedents on the left and consequents on the right). Both external and internal sources are indicated on the left. All blocks indicate states, while lines indicate process. With regard to relationship, the lines all indicate two-way flows, for diffusion to occur there must be interaction between the two cultural systems. The scheme also considers both cultural-behavioural interactions and the mutual behavioural interactions (influence) of outsiders and indigeneous residents. These four sources that are considered to be antecedent then enter into an interaction with the consequents; however the major direction of influence is stronger from antecedent to consequent than the return influence. Finally, a mutual interaction is envisaged between the changed socio-cultural system and the changed behaviours. And with respect to content, since the primary interest is the individual, a number of variables are implicated : ecological setting and economic resources, political development (among internal sources), and technological and formal education (among external sources). Other implicated features are the relationships between the two cultures, particularly the demographic factors of migration and population dominance, and the nature of intergroup relations (whether plural or monistic and whether positive or hostile).

At the level of individual, a number of psychological variables (described hereafter) are important. Among the external antecedents are the behaviours and beliefs brought by educators, missionaries, traders, and colonial officials. Among the internal antecedents are such psychological features as attitude toward change, achievement cognition, orientation goals and means, and other personality characteristics. Finally, among the consequents, classes of behaviours are apparent: (i) behavioural shifts are change in behaviour toward new norms; and (ii) acculturative stress that refers to the disorganization or even disintegration of behaviour that often (but not inevitably) accompanies social and cultural change and development (Berry, 1980).

Given the strong theoretical and research ties between cross-cultural psychology and anthropology, in broad terms, there are two paradigms for research : (i) the acculturative stress (having the longest history and the greater theoretical impact); and (ii) the development (a more recent and applied) orientation (Berry, 1980).

Psychological Antecedents of Social Change and Development

Psychological studies on the antecedents to change, particularly attitudes and beliefs (Inkeles and Smith, 1974; Smith and Inkeles, 1966; Kahl, 1968, Inkeles, 1977) on the universal approach (Inkeles and Smiths, 1974) on the local exampler of universal process (Gusfield, 1967; Stephenson, 1968; Dawson, 1967, 1969); on endogeneous beliefs (Jahoda, 1961, 1962, 1968, 1970; Jones, 1977; Kahl, 1965 and Guthrie, 1970), achievement orientation : particularly need for achievement (McClelland, 1961, 1980, 1985) provide empirical bases on the differences with regard to the psychological characteristic both for within and between cultures, and ascert that these processes are the crucial factors that determine the strength of behaviour of people across cultures for development. In addition, a global personality approach concerned with the congruity between the global personality of the group being influenced and that of the group that is bearing the acculturative pressure has set in the research tradition of "culture and personality" and of "value orientation" as the antecedents of social change and development (Mead, 1956a & b; Beaglehole 1957; Hallowell, 1955; Wallace, 1951). Although their works were more specifically directed towards constructing the "modal personality" of the people they were working with, all considered the questions of social and cultural change and the likely role of the personality they were describing. A common theme was that if the personality was oriented toward change, particularly if they were congruent with the life style which they were increasingly aware, then social change would be relatively smooth and rapid. A number of field studies on a specific trait approach (Hagen, 1962; McClelland and Winter, 1969; Smith, 1974) attained a good deal of attention on personality characteristic and social change.

The basic argument of these researchers have been that in agricultural societies, economic and technological growth occurs only gradually "contact with technologically advanced society is a necessary condition" (Hagen, 1962, pp. 34-35), but it is not sufficient. Other elements, such as material resources and social and psychological factors must also be present, and must be structured in certain ways. Thus the key to growth "seems to be largely internal rather than external" (Hagen, 1962, p. 55), and the central feature lies in the concept of "traditionalism". For Hagen, in a traditional society "behaviour is governed by customs, not law". The social structure is hierarchical. The individuals position in the society is normally inherited rather than achieved. And, at least in the traditional state so far in the world's history, economic productivity is low (pp. 55-56). Such a social system is "stable because the simple folk as well as the elite accepted it (p. 71). In contrast to the pervasive authoritarian personality, Hagen defines the "innovational personality" in a term of creativity, positive attitude towards working in novel field and openness to new experience. For Hagen (p. 86) "social change will not occur without change in personality", and it is a change from the innovational that is required.

Many other cognitive aspects have been proposed as relevant to social change and development in psychology and anthropology. Perhaps the most widely accepted view was that some peoples simply were not "intelligent" enough to "progress". This view is not considered seriously any longer, and that the fallacies and ethnocentrism inherent in such a view have been sufficiently aired (Berry & Dasen, 1974; Cole & Scribner, 1974). However, to dismiss this first approach is not to dismiss the question of group difference (in contrast to deficits) in cognitive abilities and structures, or their possible roles in the process of social and cultural change and developmental. Cross-cultural studies on perception and cognition suggest that there are differences in the perceptual and cognitive skills developed by the people to meet their particular ecological and cultural problems. Studies also suggest that perceptual and cognitive skills required by the industrialised and technological societies and their life style are of a particular kind, the requirement is for relatively well developed perceptual (disembedding spatial) skills and high development of analytic cognitive skills (Deregowski, 1973; Bowd, 1974). One problem of social and cultural change from this point of view, is to achieve a match or congruence between the patterns of a skills developed and valued in traditional life (Berry, 1971, p. 143; MacArthur, 1975, p. 240). Alternatively, social and cultural change, rather than moving towards technological life, could move towards a life style that capitalizes on extent skill patterns. In the former, change would involve the deliberate alteration of a skill pattern to operate for a standard technological life style, while in the latter, change would involve the development of a new life style that better matches the present pattern of perceptual and cognitive skills. It may be recalled, as described elsewhere, after a short lived euphoria of material gain to the detriment of human and social values, the importance of socio-cultural systems and practices, the human values and the life style, as well the needs and expectations of the people and there fulfilments has acquired the focal theme for formulation of developmental programmes and strategies to fulfil the ideals of human abilities in cultural and cross-cultural contexts. While referring to these social and human values in relevance to the socio-cultural systems and practices, provokes to pin point that uniform policies and strategies (for upliftment of back trodden class of people) were planned and implemented without considering the social and cultural values, and it seems reasonable to state that different communities reacted in their own characteristic ways (may be unaware of the nature of development programmes) and hence disparity in development. This is not something novel rather follows the leads from others (Sinha & Kao, 1988; Tri et al., 1988) that indegenous models of development be encouraged.

Pareek (1968) argued that motivation is important for social change, in general, and development in particular; but achievement motivation alone cannot promote development. Two other motives as Pareek calls "*extension motivation*" and "*dependence motivation*" must also be considered. The former implies a "concern for other people or the society" and is defined as "a need to extend the self or the ego and to relate to a large group and its goals" (Pareek, 1968, pp. 118-119). This nation, however, is not dissimilar to the notion of "affirmation" concept and arguments made by de Vos (1968) for achievement. The latter is considered to be "super-ordinate", making it useful both for improving intergroup harmony, and for "sustaining continued motivation of people in development (Pareek, 1968, p. 119). However, no empirical evidence is available to either establish the existence of this motive or to evaluate its role in social change, Pareek (p. 119) argues that there is considerable historical and political evidence suggesting such a motive, this aspect still awaits scientific exploration. The latter, that is, "dependence motivation" is also thought to be important in the process of development because it is a negative factor. Pareek (1968, p. 119) defines it as "looking for direction from other sources : and may be manifested in either" excessive-dependence-seeking support and guidance, or in excessive counter-dependency-the aggressive rejection of authority. It is characterized by the lack of initiatives, avoidance of responsibility, direction seeking, seeking favour of supervisor, and over conformity.

In a statement of general paradigm, Pareek (1968, p. 121) suggests that the development (D) is positive function of achievement motivation (AM) and extension motivation (EM), reduced by the degree of dependence motivation (DM) [D -> (AMxEM - DM). To a certain extent, ascending to Berry (1980), this formulation is only a marginal extension of the views of McClelland. Both recognized that achievement motivation is not all there is to development, and both are concerned with the obvious need for entrepreneurical activity not to be limited to self-qualification (McClelland refers to it as a "concern for the common welfare of all", while Pareek calls it the "extension motivation"). Finally, "dependency motivation" once operationalized may turn out to be little more than the polar opposite of achievement motivation, and hence its presence in the paradigm is mathematically redundant. Berry (1980) makes a special mention that the elaboration of complex sets of motives are important for social change, and express the hope that Pareek (along with de Vos) must be commended and encouraged to further this line of enquiry.

This complex persuits of motives, and their relationships with other psychological (and economic) factors, has been conducted by

Himmelstrand and Okediji (1968). They pointed out that social and economic development cannot be considered simply as a function of levels of resources (including both psychological and non-psychological resources); the "structure and patterns of resources" must be examined. Basic to their argument is the notion of resource congruence, "underdevelopment, we presume, is indicated not only or even mainly by low levels of resources but rather by incongruous or imbalanced resource structures" (p. 26). At three levels of analysis (individual, group and societal), they examine some possible kinds of incongruence.

The vast arena of research on achievement motivation turns to the arguments of Dawson (1973), in the theory to account for the cross-cultural variation in "modern attitude". For Dawson (1973) not only attitudes, but also achievement motivation and "potential for economic development" will vary according to exploitive pattern or sustenance base of a society; this relation is mediated by the variation in authority system and socialization emphasis typically developed by hunting as opposed to agricultural societies. In a test of his proposal he found that "achievement independence" but not "achievement conformity" differed significantly between a group of Alaskan Eskimo and Japanese High School and University Students. Japanese scored high than Alaskan Eskimos. However, this observation is opposite to that which might be expected on the basis of McClelland's (1961) review of the contribution of socialization to group and individual differences; in his view the more achievement-oriented and self-reliant socialization emphasis of some societies should lead to higher, not lower, achievement motivation. Studies, in general, provide empirical foundations for generalization that higher level of achievement orientation - the achievement means and achievement goals serves the bases for group differences in achievement motivation and "potential for economic development". The motivational components pooled with the learning paradigms (models) are desirable at this juncture to make lucid the nature of how motives, skills and values determine what people do (McClelland, 1985).

Psychologists have long realized that if they want to know how well something will be done, irrespective of the phylogenic series, it is important to know how motivation and skill are involved. Hull (1943) formalized this relationship in his well known equation, sEr=DX sHr, in which excitatory potential (sEr) or the tendency to make a response is a function of habit strength (sHr) x drive strength (D). According to him, habit strength was the amount of reinforced practice that an organism has in making a response previously, or in more general terms the skill it has acquired in making the response. Later, Hull (1952) added the third variable to this equation to take into account the effect of incentive value on performance. The formula was expanded, again multiplicatively, to read as follows :

$$sEr = D \times sHr \times K \text{ (for incentive).}$$

Spence (1956) argued that incentive value (k) was important. However, he felt that it should not be multiplied with other determinants but should be added to drive strength and sum of this two should be multiplied by this strength. In this formulation, the presence of either drive or incentive would lead to some behaviour if any habit strength existed, whereas in Hull's formulation if either incentive or drive was zero there would be no tendency to add. Atkinson (1964) reviewed the relationship of the Hull-Spence equations to Lewinian theory, Tolman's expectancy theory, and decision theory and arrived at a formula very similar to Hull's except that the variables in it were defined in cognitive terms and were operationalized in measures obtained from human subjects rather than animals. His initial formula, as it applies to an achieving tendency, reads as follows :

$$Ts = Ms \times Ps \times INs.$$

The tendency to achieve success (Ts) is a multiplicative functions of motive to achieve success (Ms), expectancy or probably of success (Ps), and incentive value of success (INs). In referring to Ts (Atkinson, 1964; Weiner, 1980) dropped Hull's term of excitatory potential and employed the term motivation to describe the end product of all the determinants of actions. That is, anything that influence the tendency to respond was considered motivational. Thus, the term of motivation became equivalent to determination and did not have its original restricted meanings. Latest references on motivation (Franken, 1982; Weiner, 1980) include discussion of any variables (including skills, expectancy, incentives, and coping mechanism) that affect the strength of impulse to add. However, it is still awaited for a term that refers only to arouse motive state such as hunger or need for achievement. According to McClelland (1985) it seem less confusing to use the term motivation to refer such state and to use a more general term such as impulse to respond to refer the final product of all the determinants of behaviour.

Psychologists such as Spence (1956) and Schachter and Singer (1962) detached the notion of drive from any distinctive psychological accompaniments, which made it clear to think of motivation in purely cognitive terms. Atkinson and Feather (1966) moved further in direction by arguing that the term Ms x INs defined valence of success or the attractiveness of success. In this terms, Ms measured by Atkinson by the Thematic Appreception Test (TAT) or Achievement scores, because simply a measure of individual differences in the personal evaluation of succeeding at a class of activity in which the incentive value of evaluating performance in terms of a standard of excellence was involved. Stated otherwise, the notion that drives or motives has uniquely different physiological and effective basis was replaced by a purely cognitive conception of motives as the product of expectancies and values, accompanied by a state of physiological arousal. As the attempt of review is confined that how motives, skills, and values determined what people do, the theoretical implications of these term are not felt very much desirable, however, it may be plainly stated that the cognitive implications of his theory has drop the implications of TAT (Feather, 1982; Weiner, 1980) in favour of other measures that rely on more direct methods of arousing individual differences in the strength of valence. The reasons were many.

In the first place there is evidence that specific physiological processes are associated not only with biological drives like hunger (Mayer, 1955) but also with social motives like the need for power as measured in the TAT (McClelland et al; 1980a & b). Parallel cognitive representations of this motives in self-report do not have such correlate. In the second place, motive as a measure in the TAT have more promise for predicting long-term operant trends in life than cognitively guided self-report (McClelland, 1980). Third, taking into account both motives in Hull's original sense and values as represented by the cognitive revolution in motivational theory certainly helps improve our ability to predict behaviour (McClelland, 1985).

Ms, the motive to achieve success, was operationalized as the n-Achievement score obtained from coding TAT stories, although Atkinson's general model (1958) was supposed to apply to any motive disposition scored in this way. That is, the tendency to seek any goal was conceived as the product of the motive for that goal, as measured in the TAT, times the expectancy of achieving it times the incentive value of the class of activities defining the goal. For the sake of simplicity in exposition, Atkinson's equation will be used to refer to the generic model of how motives, expectancies and incentives combine to produce goal seeking. That is, the s in the equation will be taken to refer to success in seeking any goal rather than just to success in achievement situations, which is the way Atkinson (1964) defineds.

Later, Atkinson and Birch (1978) shifted away from what they called the traditional "episodic" view of behaviour to a view that emphasized shifts in the stream of behaviour. This focused greater attention on the percentage of time spent in various activities rather than on more traditional measures of the strength of a response to a stimulus such as choice, latency, and resistance to extinction. In this respect, Atkinson and Birch agreed with Skinner's (1966) suggestion that the probability of the occurrence of an operant response strength what is commonly called "purpose". That is, the frequency with which a rat presses a bar to get food "when no correlated stimulus can be deteched" (Skinner, 1966, p. 21) could easily be seen as representing the strength of its purpose (or drive or motive) in seeking food as opposed to doing other things (scratching itself, sniffing etc.). Atkinson (1981) went further and suggested that the frequency orientation implied that the frequency with which achievement thoughts appeared in a TAT (or presumably any achievement related activity) was a function not of the absolute strength of the achievement motive but of its strength relative to all the other motives operating in the situation.

Ps, or probability of success in attaining a goal, might be considered to be the cognitive equivalent of Hull's habit strength because the probability of success obviously varies with habit strength or skill. But Ps has additional meaning because it is determined not only by actual skill but also by the individual's beliefs about the efficacy of making a response that may be somewhat independent of the individual's skill in making it. Two types of such beliefs have been studied extensively. One type has to do with the efficacy of effort in bringing about a consequence through a particular response in a given situation (Weiner, 1980). The other type has to do with the generalized confidence (or lack of it as in learned helplessness, Seligman, 1975) a person has that he or she can bring about outcomes through instrumental activities of any kind. Here, work by deCharms (1976) on the effects of helping people to feel like "origins" (originators) rather than pawns, or Bandura's work on self-efficacy training (1982) deserve mention. Along with actual skill, such beliefs enter in to determine the Ps variable.

The INs, or incentive value of success in attaining a goal, was originally conceptualized by Atkinson (1957) in a limited way to explain risk-taking behaviour in an achievement situation. He defined the incentive value of success in an achievement situation as 1 - Ps, meaning that the more difficult the task (or the less the probability of succeeding at it), the greater the reward value of succeeding in performing it. If 1 - Ps is substituted in the general equation for predicting an achieving tendency (Ts), it multiplies with other variables to produce the strongest tendency to achieve success when Ps = .50, which corresponds fairly well with actual preferences for tasks varying in difficulty. But in terms of Atkinson's general model, incentive value will be different for different motives, although he did not give much attention to how such incentives should be measured.

Even within the context of achievement motive theory, Parsons and Goff (1978), Maehr (1974), Maehr and Kleiber (1981), and others have asked what kinds of achievement or success are valued by older people, by people from cultures, or by women, and they have found evidence that far more than difficulty of performance is involved in determining such incentive values. At this point, three conclusions seems justified about the extensive controversy that has developed over the issue of how to define the achievement incentive (McClelland, 1985).

1. There is ample evidence that the moderate challenge incentive is crucial for individuals high in n-Achievement; they will work harder when this incentive is present than when it is not present; that is, when tasks are too easy or too hard (Atkinson, 1958; Clark & McClelland, 1956; French, 1955). Nor will subjects high in n-Achievement work harder when other incentives like getting time off from work are present (French, 1955).

2. The moderate challenge incentive seems to affect the performance of all people to some degree, regardless of their level of N-Achievement (Atkinson, 1958).

3. Many other incentives affect performance such as time off from work (French, 1955), cooperative rather than competitive work (Gallimore, 1981), or career or social orientation in women (French & Lesser, 1964). So when Maehr (1974) criticized achievement motivation theory for being ethnocentric and individualistically oriented, he was right only to the very limited extent that the challenge incentive is considered the only way to define INs. Obviously, many other ways of defining such incentives exist. Even in the case of the challenge incentive, other values determine the areas of behaviour in which such challenge are sought (French & Lesser, 1964).

What this means is that other more direct ways should be found for measuring the INs variable. The cognitive theorists have made their greatest contribution in this area because they have focused on asking subjects in a variety of ways what their goals are, how important achievement is to them, what defines success for them, and so on. Summary scores based on such value attitude surveys may be regarded as direct measures of the INs variable. A generation ago, deCharms, Morrison, Reitman, and McClelland (1955) demonstrated that the behavioural correlates of such a measure of achievement values (which they labelled v-Achievement) were quite different from the behavioural correlates of the TAT n-Achievement variable. For instance, n-Achievement was significantly related to better performance on an anagrams test and to better recall of the achievement content of stories, whereas v-Achievement was not. Subjects high in v-Achievement, on the other hand, were more impressed by expert judgement, whereas those high in n-Achievement were not. In other words, strong achievement values affect cognitive judgements, as they should if they are tapping INs.

McClelland (1980) has summarized the empirical evidence demonstrating that v-Achievement and n-Achievement measures are not tapping the same variable and that in general TAT motive measures are better at predicting long-term operant trends in action, whereas value-attitude measures are better at predicting choices, attributions, and other such cognitively guided behaviour. In the light of this evidence, it is misleading to consider self-report value attitude measures to be measures of human motive strength as Weiner (1980) did, for example, in considering scores from the Mehrabian (1969) achievement attitude survey to be a measure of achievement motivation. This not only makes it difficult to collate findings from two different, essentially unrelated measures of the achievement motive, but it also collapses the motive and incentive variables that, in Hull's and Atkinson's formulations, are independent determinants of action. The positive contribution of measures like Mehrabian's is precisely that it helps to define the INs variable in the equation for the dynamics of action.

To sum up, psychological studies on attitudes and beliefs, social values and work ethics, authoritative socialization, specific, personality trait approach, perceptual-cognitive profiles, the motivational characteristics - achievement means, achievement cognitions and achievement goals - need hierarchy of people, competition tolerance, flexibility and change proneness with special reference to socio-cultural change and development - as measures at individual level - provide ample evidence for differences in individuals - the social being (the man) - in culture specific and cross-cultural perspectives. Viewed in these perspectives, the development strategies be endogeneous, firstly, must be based on the actual circumstances of societies and needs and aspirations of their populations and, secondly, on the existing and potential resources whether human, material, technical or financial, which any society may possess, while taking into account the numerous constraints that are inherent in such circumstances. Every society must seek its own type or style of development in accordance with the features of its own culture and patterns of thought and action. Every specific society has its own patterns and models of development. And there is no such things as a single, uniform development model : experience acquired in recent decades has shown that no development model can be viewed as being universal or universally transposable in terms either of space or time.

Additionally, the knowledge of ecological psychology suggest that differentials in the environment (both the geographical and physical as well as the psychological and social) affects differently the development of cognitive skills differently, and that of the many effects produced by deprivation in the environment, the need security is perhaps the only one which seems to constitute the care of healthy functioning of the individual, and this need security becomes the corn stone of the edifice of self actualization. Studies suggest that those individuals who struggle at the basic levels of their need cannot rise to the level of creative needs, and deprivations induces deficiency in need oriented behaviours like insecurity, anxiety (and to a greater extent pathology), and only when the basic needs are fulfilled, an individual can rise on the steps of need hierarchy ladder. Thus, in order to achieve authentic development, the utmost felt need is to find congruence between the psychological factors with reference to the structural properties of the environment : both geographical and physical as well as the psychological and social. The psycho-social and cultural dimensions pooled with the technological considerations (with their limitations and imperfections) in culture-specific and cross-cultural perspectives are desirable for endogeneous development, with its own project of civilization as its goals and man himself as its primary concern.

The theorizations made with regard to psychosocial and cultural factors of social change and development, by implication, provide empirical foundations to suggest that human factors be thoroughly viewed before implementation of any model, at least in the first phase, for successful adoptation of the technology and successfulness of development programmes so that the changed peoples (the beneficiaries of the development programmes) would serve as the socializing agents for others for adoptation. Studies on these counts are scanty, however, findings of such studies would provide empirical bases to formulate strategies for cognitive and behaviour modification of peoples and community orientation programme to facilitate patterns relating to social and agroforestry (and for any development programme) for fruitful achievements.

Acknowledgements

The author wish to thank Dr. L.N. Jha, Professor of Forestry, North-Eastern Hill University for his consistent encouragements and helping comments. Thanks are also acknowledged to Miss Lalramengi Ralte, Irene Zohlimpuii Chongthu, and Sangliankhumi Hmar, the Post Graduate students of the department, for their manner of service in reviewing the topic, and arranging the references for presentation.

REFERENCES

Atkinson, J.W. (1957). Motivational determinants of risk-taking behaviour. *Psychological Review, 64,* 359-372.

Atkinson, J.W. (1958). Towards experimental analysis of human motivation in terms of motives, expectancies, and incentives. In Atkinson, J.W. (Ed.) *Motives in Fantasy, Action and Society* (pp. 288-305). Princeton, NJ : Van Nostrand.

Atkinson, J.W. (1964). *An Introduction to Motivation.* New York : Van Nostrand.

Atkinson, J.W. (1981). Thematic apperceptive measurement of motivation in 1950 and 1980. In G. D'Ydewalle and W. Lens (Eds.) *Cognition in Human Motivation and Learning* (pp. 159-198). Hillslale, NJ : Erlbaum.

Atkinson, J.W., & Birch, D. (1978). *Introduction to Motivation* (2nd Ed.). New York : Van Nostrand.

Atkinson, J.W., & Feather, N.T. (1966). *A Theory of Achievement Motivation.* New York : Wiley.

Bandura, A. (1982). Self-efficacy mechanism in human agency. *American Psychologist, 37,* 122-147.

Beaglehole, E. (1957). *Social Change in the South Pacific*. New York : MacMillan.

Berry, J.W. (1971). Psychological research in the North. *Anthropologica, 13*, 143-57.

Berry, J.W. (1980). Social and cultural change. In H.C. Triandis and R.W. Bristin (Eds.). *Handbook of Cross-Cultural Psychology & Social Psychology* (Vol. 5). Allyn and Bacon, Inc. Boston...Toronto.

Berry, J.W., and Dasen, P. (Eds.) (1974). *Cultural and Cognition*. London : Methuen.

Bowd, A. (1974). Practical abilities of Indians and Eskimos. *Canadian Psychologist, 15*, 281-90.

Campbell, A., and Converse, P. (Eds.), (1972). *The Human Meaning of Social Change*. New York : Russell Sage Foundation.

Capra, F. (1982). *The Turning Point*. New York : Simon and Schuster.

Clark, R.A. and McClelland, D.C. (1956). A factor analytic integration of imaginative and performance measures of the need for achievement. *Journal of General Psychology, 55*, 73-83.

Cole, M., and Scribner, S. (1974). *Culture and Thought*. New York : Wiley.

Dawson, J.L.M. (1967a). Traditional versus western attitudes in West Africa : the construction, validation and application of a measuring device. *British Journal of Social and Clinical Psychology, 6*, 81-96.

Dawson. J.L.M. (1969a). Attitudinal consistency and conflict in West Africa. *International Journal of Psychology, 4*, 39-53.

Dawson. J.L.M. (1973). Effects of ecology and subjective culture on individual traditional-modern attitude change, achievement motivation, and potential for economic development in the Japanese and Eskimo societies. *International Journal of Psychology, 8*, 215-25.

deCharms, R. (1976). *Enhancing Motivation : Change in the Class-room*. New York : Irvington.

deCharms, R., Morrison, H.W., Reitman, W.R., and McClelland, D.C. (1955). Behavioural correlates of directly and indirectly measured achievement motivation. In D.C. McClelland (Ed.), *Studies in Motivation* (pp. 414-423). New York : Appleton-Century-Crofts.

Deregowski, J.B. (1973). Industrialization of developing countries : problem of simple skills. *International Review of Applied Psychology, 22*, 77-84.

Desai, A.R. (Ed.) (1971). *Essays on Modernization of Under-developed Societies*. (2 vols.). Bombay : Thacker.

de Vos, G. (1968). Achievement and innovation in culture and personality. In E. Norbeck, D. Prince-Williams, and W. Mc Cord (Eds.), *The Study of Personality : An Interdisciplinary appraisal*. New York : Holt, Rinehart and Winston, pp. 348-70.

Feather, N.T. (Ed.) (1982). *Expectations and Actions : Expectancy Value Models in Psychology*. Hillsdale, NJ : Erlbaum.

Franken, R.E. (1982). *Human Motivation*. Montercy, CA : Brooks/Cole.

French, E.G. (1955). Some characteristics of achievement motivation. *Journal of Experimental Psychology, 50*, 232-236.

French, E.G., and Lesser, G.S. (1964). Some characteristics of the achievement motive in women. *Journal of Abnormal and Social Psychology, 68*, 119-128.

Gallimore, R. (1981). Affiliation, social context, industriousness and achievement. In R.H. Munroe and B.B. Whiting (Eds.), *Handbook of Cross-Cultural Human Development*. New York : Garland STPM.

Gusfield, J.R. (1967). Tradition and modernity : misplaced polarities in the study of social change. *American Journal of Sociology, 72*, 351-62.

Guthrie, G. (1970). *The Psychology of Modernization in the Rural Philippines*. Quezon City : Atenco de Manila University Press.

Hagen, E. (1962). *On the Theory of Social Change*. Homewood, Ill : Dorsey Press.

Hallowell, A.I. (1955). *Cultural and Experience*. Philadelphia : University of Pennsylvania Press.

Himmelstrand, U., and Okediji, F.O. (1968). Social structure and motivational tuning in social and economic development. *Journal of Social Issues*, 24, 25-42.

Hull, C.L. (1943). *Principles of Behaviour.* New York : Appleton-Century-Crofts.

Hull, C.L. (1952). *A Behaviour System.* New Haven : Yale University Press.

Inkeles, A., and Smith, D. (1974). *Becoming Modern.* Cambridge, Mass : Harvard University Press.

Inkeles, A. (1975). Applying cross-cultural psychology to the Third World. In J.W. Berry and W.J. Lonner (Eds.), *Applied Cross-Cultural Psychology,* Amsterdam : Swets and Zeitlinger, pp. 3-7.

Inkeles, A. (1977). Understanding and misunderstanding individual modernity. *Journal of Cross-Cultural Psychology, 8,* 135-76.

Jahoda, G. (1961). Aspects of westernization, I. *British Journal of Sociology, 12,* 375-86.

Jahoda, G. (1962). Aspects of westernization, II. British *Journal of Sociology, 13,* 43-56.

Jahoda, G. (1968). Some research problems in African Education. *Journal of Social Issues, 24,* 161-75.

Jahoda, G. (1970). Supernatural beliefs and changing cognitive structures among Ghanaian University students. *Journal of Cross-Cultural Psychology, 1,* 115-30.

Jones, P. (1977). The validity of traditional - modern attitude measures. *Journal of Cross-Cultural Psychology, 8,* 207-40.

Kahl, J. (1968). *The Measurement of Modernism : A study of Values in Brazil and Mexico.* Austin, Texas : University of Texas Press.

Kavolis, V. (1970). Post-modern man : psychocultural responses to social trends. *Social Problems, 17,* 435-48.

Kelman, H.C., and Warwick, D.P. (1973). Bridging micro and macro approaches to social change : a social psychological perspective. In G. Zaltman (Ed.), *Processes and Phenomena of Social Change.* New York : Wiley pp. 13-449.

Kunkel J. (1966). *Society and Economic Growth : A Behavioural Perspective of Social Change.* London : Oxford University Press.

Lewis, W.A. (1955). *Theory of Ecoomic Growth.* London : Allen and Unwin.

Mac Arthur, R.S. (1975). Differential ability patterns : Invit, 'Nsenga and Canadian Whites. In J.W. Berry and W.J. Looner (Eds.), *Applied Cross-Cultural Psychology*. Amsterdam : Swets and Zeitlinger.

Maehr, M.L. (1974). Culture and achievement motivation. *American Psychologist, 29,* 887-896.

Maehr, M.L., and Kleiber, D.A. (1981). The graying of achievement motivation. *American Psychologist, 36,* 787-793.

Mayer, J. (1955). Regulation of energy intake and body weight : the glucostatic theory and the lipostatic hypothesis. *Annals of the New York Academy of Science, 63,* 15-43.

Mazrui, A.A. (1968). From social Darwinism to current theories of modernization : a tradition of analysis. *World Politics, 21,* 69-83.

McClelland, D.C. (1961). *The Achieving Society*. New York : Van Nostrand.

McClelland, D.C. (1980). Motive dispositions : the merits of operant and respondent measures. In L. Wheeler (Ed.), *Review of Personality and Social Psychology* (Vol. I). Beverly Hills, CA : Sage.

McClelland, D.C. (1984). *On the Two Psychologies of Love*. Unpublished manuscript, Department of Psychology and Social Relations, Harvard University.

McClelland, D.C. (1985). *Human Motivation*. Glen View, IL : Scott, Foresman.

McClelland, D.C; Davidson, R.J., Floor, E. and Saron, C. (1980a). Stressed power motivation, sympathetic activation, immune function and illness. *Journal of Human Stress, 6(2),* 11-19.

McClelland, D.C., Davidson, R.J., Saron, C., and Floor, E. (1980b). The need for power, brain norepinephrine turnover and learning. *Biological Psychology, 10,* 93-102.

McClelland, D.C. and Winter, D.G. (1969). *Motivating Economic Development*. New York : Free Press.

Mead, M. (Ed.) (1956a). *New Lives for Old*. New York : Morrow.

Mead, M. (1956b). The implications of culture change for personality development. In D.G. Haring (Ed.), *Personal Character and Cultural Milieu*. Syracuse, N.Y.: University of Syracuse Press.

Mehrabian, A. (1969). Measures of achieving tendency. *Educational and Psychological Measurements, 29*, 445-451.

Nisbet, R. (1971). Ethnocentrism and the comparative method. In A.R. Desai (Ed.), *Essays on Modernization of Underdeveloped Societies*, Vol. I. Bombay : Thacker, pp. 95-114.

Omvedt, G. (1971). Modernization theories : the ideology of empire ? In A.R. Desai (Ed.), *Essays on Modernization of Underdeveloped Societies*, Vol. I Bombay : Thacker, pp. 119-37.

Pareek, U. (1968). A motivational paradigm of development. *Journal of Social Issues, 24*, 115-22.

Park, R.E. (1928). Human migration and the marginal man. *American Journal of Sociology, 33*, 881-93.

Parsons, J.E., and Goff, S.B. (1978). Achievement motivation and values : An alternative perspective. In L.J. Fyans, Jr. (Ed.), *Achievement Motivation : Recent Trends in Theory and Research* (pp. 349-373). New York : Plenum Press.

Rostow, W.W. (1952). *The Process of Economic Growth*. New York : Narton.

Schachter, S. and Singer, J.E. (1962) Cognitive, Social and Physiological determinants of emotional states. *Psychological Review, 69*, 379-399.

Schwendler, W. (1984). UNESCO's project on the exchange of knowledge for endogeneous development. *International Journal of Psychology, 19*, 3-15.

Seligman, M.E.P. (1975). *Helplessness : On Depression, Development, and Death*. San Francisco : Freeman.

Sinha, D. and Kao, H.S.R. (1988). *Social Values and Development : Asian Perspectives*. Sage Publications. New Delhi, London.

Skinner, B.F. (1966). *The Behaviour of Organisms*. Englewood Cliffs, NJ : Prentice Hall.

Smith, D. and Inkeles, A. (1966). The OM Scale : A comparative socio-psychological measure of individual modernity. *Sociometry, 29*, 353-77.

Spence, K.W. (1956). *Behaviour Theory and Conditioning*. New Haven, CT : Yale University Press.

Stephenson, J. (1968). Is everyone going modern ? a critique and a suggestion for measuring modernism. *American Journal of Sociology, 74*, 265-75.

Tri, H.C., Khoi, L.T., Colin, R., Ceestem, Yuan-Zheng, L. (1986). *Strategies for Endogenous Development*. Oxford and IBH Publishing Co. Pvt. Ltd., New Delhi, Calcutta.

Triandis, H.C. (1971). Some psychological dimensions of modernization. *Proceedings of the 17th Internation Congress of Applied Psychology*. Bruxelles, Vol. 2(12), 57-65.

Triandis, H.C. (1973). Subjective culture and economic development. *International Journal of Psychology, 8(3)*, 163-180.

Wallace, A.F.C. (1956). Revitalization movements : some theoretical considerations for their comparative study. *American Anthropologist, 58*, 264-81.

Wallace, A.F.C. (1951). Some psychological determinants of cultural change in an Iroquoian community. In W.N. Fenton (Ed.), *Symposium on Local Diversity in Iroquois Culture*. Washington, D.C.: Bureau of American Ethnology, Bulletin, 149.

Weber, M. (1958). *The Religious of India : The Sociology of* Hinduism and Buddhism. Glencoe : Free Press.

Weiner, M. (Ed.) (1966). *Modernization : The Dynamics of Growth*. New York : Basic.

Weiner, B. (1980). *Human Motivation*. New York : Holt, Rinehart, and Winston.

Index

General

Botanical Names

Zoological Names